THE RIVER OF NO RETURN

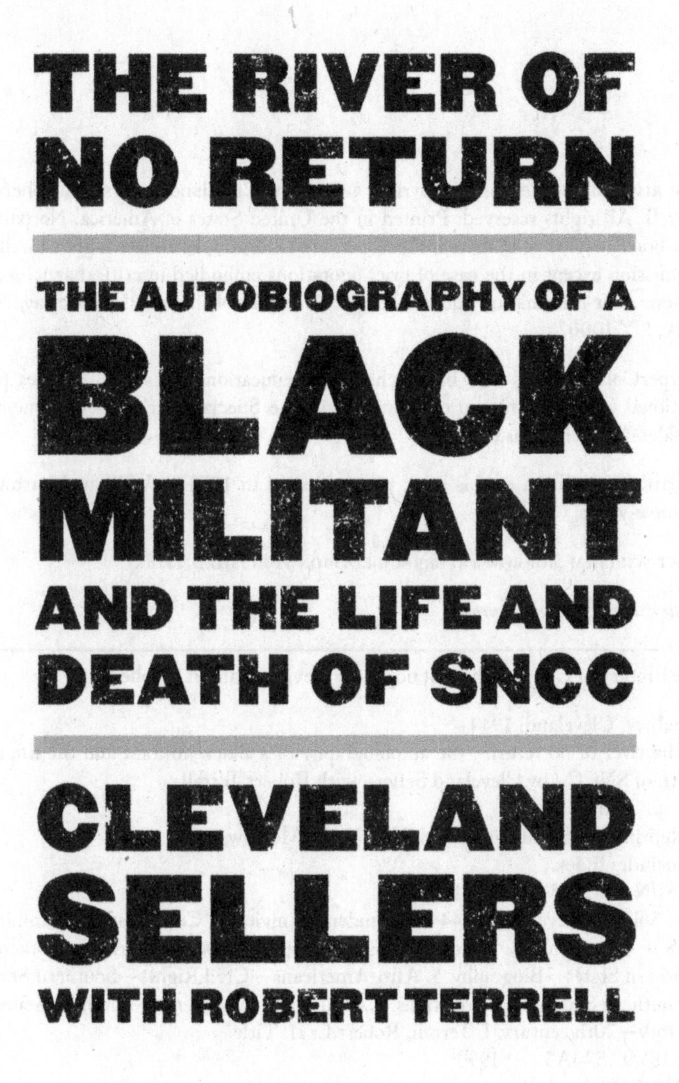

wm

WILLIAM MORROW

An Imprint of HarperCollins*Publishers*

HarperCollins books may be purchased for educational, business, or sales promotional use. For information please email the Special Markets Department at SPsales@harpercollins.com.

A hardcover edition of this book was published in 1973 by William Morrow & Company, Inc.

FIRST WILLIAM MORROW PAPERBACK EDITION PUBLISHED 2018.

Designed by Diahann Sturge

The Library of Congress has catalogued a previous edition as follows:

Sellers, Cleveland, 1944–
The river of no return : the autobiography of a Black militant and the life and death of SNCC / by Cleveland Sellers, with Robert Terrell
p. cm.
Reprint. Originally published: New York : Morrow, 1973.
Includes index.
ISBN 0–87805–474-X (alk. paper)
1. Sellers, Cleveland, 1944–. 2. Student Nonviolent Coordinating Committee (U.S.)—Biography. 3. Afro-Americans—Biography. 4. Civil rights workers—Southern States—Biography. 5. Afro-Americans—Civil Rights—Southern States. 6. Southern States—Race relations. 7. Civil rights movements—Southern States—History—20th century. I. Terrell, Robert L. II. Title.
E185.97.S44A3 1990
322.4'4'092—dc20
[B] 90–12847
CIP

ISBN 978-0-06-282431-8 (pbk.)

HB 12.30.2022

CONTENTS

THE RIVER OF NO RETURN

CHAPTER 1

Growing Up in Denmark

I WAS BORN on November 8, 1944, in a central South Carolina town called Denmark. It was divided into nearly equal sections by rusting, seldom-used railroad tracks. In a larger, more prosperous town, the sections created by the tracks would probably have been individually populated by distinct social classes. This was not the case in Denmark—it was too small.

But like every other Southern town, Denmark did have strict social divisions—they were racial. Its three thousand residents, half of whom were black, lived for all practical purposes in different worlds. I know a great deal, almost everything about black Denmark. And almost nothing about white Denmark: the divisions were that strict.

Black Denmark is family-oriented and my earliest memories are of my family. My father, a dark almost black man with large wrists, strong hands and tough palms, is the central figure in the recollections of my childhood. He was always at work when I got out of bed in the mornings. He seemed to work all the time.

Sometimes he returned home in time to play with me before

I went to bed. On those rare occasions, I always marveled at his fingers and the skin on his face. His fingers were gnarled from long years of hard work. And his face was finely creased from the wind and sun.

My mother is a smooth, graceful woman with honey-brown complexion and delicate limbs. Her hands are strong too, but the skin on them is soft and soothing. My earliest memories of my mother are related to her jobs, too. From as far back as I can remember, she had been employed as a dietician and English teacher at Denmark's South Carolina Area Trade School (SCAT). Like my father, she also left for work early in the mornings. I can remember very few times when my sister and I got out of bed and my mother was still home.

My sister Gwendolyn is two years older than I. Because we were alone a lot when we were young, we were very close. We enjoyed each other's company and spent many mornings talking excitedly about the things we intended to do before our parents returned in the evening.

Even though our parents worked most of our waking hours, my sister and I never felt neglected. Our parents were totally devoted to us and gave us almost everything we wanted. When my father saw me get excited about a picture of a horse in a book, he went out and bought me a real horse. When he discovered that we liked to swim, he built a swimming pool in the back yard.

We always had a nice home, more than enough food to eat, decent clothing and lots of toys. My father, who came from a large family, was no stranger to poverty. He was determined that Gwendolyn and I would have all those things which he did not have when he was growing up.

My father believed that hard work would solve almost any problem. His theories of success were almost identical with those of Booker T. Washington, whom he revered. He took great pride in the financial "cushion" which he was creating for me to inherit when I grew up. If the people next door got up and began working at five o'clock in the morning, my father would get up at four o'clock. That's the kind of man he was.

When I was very young, my father was a part-time farmer. He also owned a restaurant and a taxi. He drove the taxi when he was not busy at the restaurant. The farm, which encompassed a few acres, was one of his many hustles. He kept a few chickens, pigs and cattle. He planted corn and soybeans for the animals and to feed us. He worked at the farm during his "spare time." He was not interested in becoming a full-time farmer. To him, the farm was just another means of making a few dollars and getting a little further ahead.

By the time I reached my teens, my father owned more than twenty houses. He built most of them himself. They weren't elaborate—just small cement-block structures with two bedrooms, a living room and a kitchen. He built them with money he saved from his other businesses. He began with four, saved more money, and built four more. He was not money hungry, just obsessed with security.

My mother was much less concerned with security than my father. The differences in their attitudes were probably related to their different family and educational backgrounds. My mother's family was wealthy compared to my father's. And unlike my father, who received his high school diploma in the army, my mother went to college and earned a degree in home economics.

The differing attitudes that my parents had toward security did not harm their marriage, however. It was always understood that my father was the head of the house, and, during disagreements, "the bull of the woods."

I became aware of poverty at the age of ten; at about the same time I began working in my father's restaurant. Before then I never really thought about poverty. I knew that there were people living in Denmark who did not have as much as we did, but I never thought of them as being poor. As far as I was concerned, they were just temporarily unlucky.

The event that really opened my eyes to poverty and the kinds of things it can force people to do took place one afternoon in grammar school. We had just finished eating lunch. As the teacher passed around the trash can so we could dispose of our lunch bags, I saw one of my friends do a strange thing. Reaching furtively into the garbage, he removed a half-eaten sandwich. No one saw him do it but me and I didn't say anything. When the trash can reached the next row, he began to eat the soggy sandwich. I watched him do this every day for a week.

I knew that he would be embarrassed if I asked him why he ate out of the trash can; so I went to my mother. She told me that his parents were poor and that they probably didn't have enough money to provide him with lunches.

"Why do people have to go hungry?" I asked. "The stores downtown are full of food."

She smiled and tried to tell me why. Her explanation still did not make sense to me. Finally, I asked her if I could take an extra sandwich each day for my friend. She said yes. My friend was embarrassed when I began to offer him sandwiches, but he took them anyway. Although we got to be very close during the remainder of the school year, I never told him that I had seen him eating out of the trash can.

My father's restaurant was one of the places where Denmark's blacks got together to socialize and have a few beers. Most of the people who frequented the restaurant were very poor. Quite a few of them were unemployed millhands with large families and slim prospects of ever getting adequate jobs.

They were the kind of people who coin phrases like, "I been down so long that down don' bother me!"

Before I began working at the restaurant, I would frequently get together with a few of my friends and visit Denmark's largest black ghetto, which was referred to as the area "across the tracks." The houses in this section of town were constructed of rough-hewn planks, propped up on bits of cement blocks. Some did not have porches. Most had never been painted and only a few had inside plumbing.

The people who lived there were desperately poor. They worked long, hard hours from Monday to Friday and on weekends drank large quantities of "white lightnin'" and "store-bought alcohol" in an attempt to drown their frustrations.

The alcohol would lower their inhibitions and release pent-up aggressions. Each weekend the area would be disrupted by terrible fights. Young and innocent, my friends and I went there to see the people fight and destroy each other. I witnessed numerous broken-bottle brawls, knife fights and shootings.

After I began working in my father's restaurant, I became painfully aware of the fact that my family was substantially better off than the people across the tracks. I also became aware of my father's blustery attitude toward poor people and poverty. I had heard him say on many occasions that anybody who really wanted to work did not have to be poor. Actually his bark was much worse than his bite. He always bent over backward to help the people who had credit at the restaurant. He was also very lenient with the people who rented houses from him. Two or three dollars and almost any excuse would make him forget about the rent until next month.

Although my mother was raised a Methodist and my father a Baptist, we attended the Episcopalian church at Voorhees Junior College. I don't know the exact reason why. It was probably because the Episcopalian church was *the place* to worship for Denmark's up and coming black middle class. In any event, I didn't mind. I even volunteered to be an acolyte. I liked the responsibility and the minister, Father Henry Grant. He understood young people.

Although we attended church every Sunday, my parents were not fanatically religious, like many Southern blacks. As far as they were concerned, regular church attendance was one of their community responsibilities. This was especially true of my father. He was a leader in the black community and believed that community leaders should give active support to the church. He frequently volunteered to serve on church committees and always contributed generously to the collection plate.

Sometimes my father would go down to the church and cut the grass. He didn't do it because he felt a religious responsibility. He did it because it was something which had to be done and he was one of the persons who was supposed to do it.

There was no contact between Denmark's black and white churches. The blacks attended their churches and the whites theirs. The fact that the churches were segregated did not bother anyone. We had no interest in the white churches. They were not relevant to our lifestyles or our needs.

I was almost a teen-ager before I really became aware of white people. I guess I always knew that white people existed, but I had very little real contact with them before my teens. Everything I did took place in a black setting. I lived in a black neighborhood. When I walked to school each morning, I passed black homes and black businesses. I attended all-black schools staffed by black administrators and black teachers. My mother taught at an all-black school and almost all my father's business dealings were with black people. No white bill collectors or insurance men ever came to our house.

Such an environment has its virtues. In an all-black community, young people are less likely to develop negative self-concepts. Because I saw black people competently performing all manner of tasks all the time, I never had any reason to question their abilities. Our parents never worried about our ability to do outstanding schoolwork. They told my sister and me what they expected us to accomplish and we tried not to disappoint them. On those rare occasions when one of us brought home a poor report card, they demanded that we tell them why. When our answers didn't satisfy them, they went to see our teachers.

This positive attitude on the part of my parents and most of the other black people I encountered during my early years engendered a great deal of self-confidence. From a very early age, I was always convinced that I could succeed whenever I really applied myself. As far as I was concerned, white people did not constitute a threat or deterrent to anything I wanted to be or accomplish.

All this should not be interpreted to mean that I was oblivious to whites. We did have some contact with them. The important thing, however, is that our contact was not fraught with tension and conflict. We never had any trouble when we went downtown, even though whites owned most of the businesses. The proprietors were always cordial; whites we met in the stores and on the streets were civil and courteous.

As far as race relations were concerned, my attitude toward Denmark's segregated movie theater was typical. I began attending the theater when I was very young. There wasn't much to do in town and it was a welcome diversion.

We always entered the side door of the theater, the one reserved for blacks, and invariably sat in the balcony, thus segregated from the whites. I never really thought about the reasons why.

All the guys in my group would go to the theater every Saturday afternoon. We sat in the same place—the front row of the balcony—and propped our feet on the banister while watching the movies. When the pictures were boring, we would throw popcorn, empty soft-drink cups and water on the whites seated below. We got a big kick out of that. The ushers would rush up to the balcony with their little red flashlights, but they never caught anyone. We always moved to another section of the balcony before they arrived.

I now realize that my unquestioning acceptance of the theater's segregationist policies grew out of a subtle, but enormously effective, conditioning process. The older people in the community, those who knew what segregation and Jim Crow were all about, taught us what we were supposed to think and how we were supposed to act. They did not teach us with words so much as they taught us with attitudes and behavior. There wasn't anything intellectual about the procedure. In fact, it was almost Pavlovian.

Denmark's blacks complied with the town's segregationist customs for the same reasons that oppressed people the world over accept and support the systems of their oppression: they had no formal power. Whites kept tight control of all the town's formal power structures. The mayor was white, all the members of the City Council were white, the police force was white, every member of the Board of Education was white and all the town's leading businesses were owned by whites.

An important thing about Denmark's power structure, however, was that whites did not flaunt their power. There were no

George Wallace or Lester Maddox types running around town yelling "Nigra" and demanding "segregation now and forever." Denmark's whites were much too sophisticated for that kind of behavior.

Whites conscientiously de-emphasized the importance of their power. Power was never publicly discussed. No one plastered banners on buildings or passed out leaflets during election campaigns. I knew we had a mayor, but I was never sure just who he was. The only thing I knew for certain was that he was white. Although the City Council held regular meetings, the time and place were never publicly announced. Blacks were free to register and vote, but very few ever did. The whites de-emphasized politics and voting so effectively that most city and county elections went unnoticed by the blacks.

Looking back, I understand a lot of things that I was oblivious to while growing up. For instance, the fanatical support that Denmark's blacks gave Voorhees Junior College and high school and the South Carolina Area Trade School (SCAT). These institutions, which were all-black, enjoyed the complete support of Denmark's blacks. In a different environment, much of the support they received would have probably been devoted to politics.

Voorhees, which is now a four-year college, was operated by the Episcopal Church. It was the best college in the area—black or white. We did not worry about racial discrimination in the municipal and county areas because we had Voorhees; and Voorhees had "the best of everything."

Denmark's South Carolina Area Trade School (SCAT) was almost as important as Voorhees. Although it was a state school, the town's blacks considered it their own. There was always a steady stream of carpenters, bricklayers, plumbers, electricians, barbers and machinists coming out of SCAT. These skilled craftsmen made it possible for Denmark's blacks to service a lot of their own needs. This contributed to our sense of proud independence.

One of the things that made Denmark's blacks feel especially close to Voorhees was the fact that all of Denmark's black high school students attended Voorhees High School. The city of Denmark operated a high school for whites but none for blacks. Despite the obvious racism involved in such an arrangement, Denmark's blacks supported it wholeheartedly.

It would be difficult to exaggerate the profound influence that Voorhees and SCAT had on black-white relations in Denmark. Their influence was felt in every area of the city's life. Although they had only about one thousand students between them, the combined spending power of the two schools and their students is one of the primary reasons why Denmark's white businessmen—and all those whites who took cues from them—were so cordial to black shoppers.

Denmark's blacks, including the administrators at Voorhees and SCAT (they were also black), were very conscious of the influence that the two schools had on the town's economy. I grew up *knowing* that as far as influence in the city of Denmark was concerned, the president of Voorhees was just as important as the mayor. Although Denmark's blacks did not always know the name of the mayor, we always knew who the presidents of Voorhees and SCAT were.

Much of the pride the students and community people had in Voorhees was manifested in support for the school's athletic teams. Football and basketball games were important community events. Everyone in the black community who was physically able attended the games. Players on the teams were community heroes.

If you played basketball or football, you did not smoke, drink or eat sweets. These things were considered harmful to good athletic conditioning, and the people in the community did not tolerate poorly conditioned athletes. It was virtually impossible for a Voorhees athlete to buy a package of cigarettes or a bottle of beer or wine from one of Denmark's black merchants. Furthermore, any athlete who was foolish enough to attempt to buy cigarettes, sweets or alcohol would be reported to the coach.

Voorhees, SCAT and athletics could not completely obliterate the specter of white racism, however. It was always there, hovering in the back of your mind like last night's bad dream. The adults tried to be relaxed about it but they couldn't. You could tell by the way they walked and talked, and the way they always managed to be completely aware of where they were and what they were doing.

The adults, my parents included, were always afraid that we young people would take white racism too lightly. They were always urging us to "be careful." They realized that we were different from them, less afraid.

Although we did not possess the same amount of fear as our parents, we did understand what white racism was and what it could do. We learned these things from a number of sources, the most important one being the grapevine: an informal, black communications network connecting state to state, town to town, group to group and person to person.

Some of the most important pieces of information passed along the grapevine were accounts of atrocities. They contained valuable survival tips for those wise enough to heed them. I can remember hearing and reflecting on such accounts from the time I was a very young boy. They almost invariably dealt with situations where black people, usually black men, were brutalized by whites.

Many of the worst accounts came from Alabama. I remember

one in particular. It was about a black man who was kidnapped and castrated by a group of marauding Ku Klux Klansmen during the late 1950's. The black man and his girlfriend were walking down a road near the outskirts of Birmingham when six Klansmen drove up and forced him into their car. He did not know his captors and they did not know him. They took the frightened man to a small, dark building where they held regular meetings and proceeded to take turns beating him.

When the Klansmen tired of beating the helpless black man, they spread-eagled him on the floor. While five of them held him, a sixth one proceeded to slice his scrotum from his body with a razor blade. They took time to douse his profusely bleeding wound with turpentine before throwing him in a ditch on the side of a country road.

When the police captured the six Klansmen, two agreed to become state's witnesses and testify against the other four. During the trial, the two informers testified that the black man was castrated because one of their members, Bart Floyd, was willing to "get nigger blood on his hands" in order to prove that he was worthy of a promotion.

The atrocity that affected me the most was Emmett Till's lynching. Emmett Till, a fourteen-year-old black youth, was lynched in Money, Mississippi, on August 28, 1955. The grapevine carried all the details of his tragic death. Emmett was spending the summer with relatives in Mississippi. Being from the North, he was not familiar with the "folkways and mores" of the area he was visiting.

One afternoon he went to a local store with a group of friends. There was an attractive young white woman in the store: Mrs. Roy Bryant, the twenty-one-year-old wife of a local redneck. Emmett whistled at her. When Mrs. Bryant's husband heard about Emmett's whistling, he became enraged. That night someone kidnapped Emmett.

Three days after the incident, a local fisherman found Emmett Till's corpse floating in the Tallahatchie River with a huge electrical fan tied to the neck. The fan had not been heavy enough to hold the bloated corpse on the bottom of the river. Emmett Till had been beaten unmercifully before he was thrown into the river. The local sheriff announced after examining the mutilated corpse that it looked like someone had used an ax on it, "the cuts were so deep." Many black newspapers and magazines carried pictures of the corpse. I can still remember them. They showed terrible gashes and tears in the flesh. It gave the appearance of a ragged, rotting sponge.

Mrs. Bryant's husband and his half brother, J. W. Milam, were arrested and charged with the killing. There was a trial. All the jurors in the court were white. Over six thousand dollars was collected from whites across the South to support the two defendants. The trial did not last very long. The two men denied that they had killed Emmett Till. The jury accepted their story and the two men were released. Blacks across the country were outraged, but powerless to do anything.

Emmett Till was only three years older than me and I identified with him. I tried to put myself in his place and imagine what he was thinking when those white men took him from his home that night. I wondered how I would have handled the situation. I read and reread the newspaper and magazine accounts. I couldn't get over the fact that the men who were accused of killing him had not been punished at all.

There was something about the cold-blooded callousness of

Emmett Till's lynching that touched everyone in the community. We had all heard atrocity accounts before, but there was something special about this one. For weeks after it happened, people continued to discuss it. It was impossible to go into a barber shop or corner grocery without hearing someone deploring Emmett Till's lynching.

We even discussed it in school. Our teachers were just as upset as we were. They did not try to distort the truth by telling us that Emmett Till's murder was an isolated event that could only have taken place in Mississippi or Alabama. Although they did not come right out and say it, we understood that our teachers held the South's racist legal system in the same low regard as we did. That's one of the good things about an all-black school. We were free to discuss many events that would have been taboo in an integrated school.

THERE WERE NO significant changes in the lifestyles of Denmark's blacks from the time of my birth until 1960, when the sit-ins began. One year was distinguished from the next by the athletic teams at Voorhees and SCAT. Some years the teams were good and other years they were very good. White control of the formal power structure went unchallenged. Although a few of Denmark's blacks belonged to the NAACP, their activities were restricted to halfhearted membership drives once or twice a year.

The grownups, black and white, considered Denmark a "good Southern town." Of course, the blacks defined *good* in different terms from the whites. When the blacks called Denmark a good Southern town, they meant that the police "minded their own business," that the Ku Klux Klan was not active and that the local whites did not maliciously flaunt their power. As far as the whites were concerned, Denmark was a good town because the blacks were not trying to change things.

Like most of Denmark's young blacks, I did not consider it a good town. By 1960—I was sixteen years old then—I was ready for change. The racism I had been conditioned to accept without notice when I was younger had begun to bother me.

The local theater was no longer *the place* to spend carefree Saturday afternoons. I understood that the theater was an integral part of a system I abhorred. It was insulting, and I realized that every time I bowed to custom by entering the side door and sitting in the balcony I was perpetuating my own degradation.

I wanted to speak out and challenge those who said that blacks were inferior. I wanted to obliterate the "white only" signs that served as ever-present reminders of our subjugation. I yearned to live in a world where I would never again be confronted with restroom signs saying, WHITE LADIES TO THE RIGHT AND COLORED WOMEN TO THE REAR.

Television was largely responsible for my new perceptions of Denmark and the South. It helped me to see our plight in new ways, helped me to understand that protest was viable, and sometimes successful. I watched television news reports of the school integration crisis in Little Rock, Arkansas, and the heroic Montgomery bus boycott. I saw Daisy Bates, Mrs. Rosa Parks and Dr. Martin Luther King, Jr., in my own living room.

I was extremely proud of them. They were my people. They were standing and fighting a common enemy. When they spoke, they said what I was thinking. When they suffered, I suffered with them. And on those rare occasions when they managed to eke out a meager victory, I rejoiced too.

I spent many long hours thinking about "the problem." I'd become particularly aware of my environment: the differences between the black and white sides of town, the numerous little indignities that blacks had to endure just to survive from day to day, the awesome poverty suffered by the vast majority of Southern blacks. My thoughts always culminated with the same vow: "I've gotta help do something about this shit!"

CHAPTER 2

The Movement Comes to Denmark

THE FIRST SIT-IN, which took place on February 1, 1960, was a beautiful surprise. The news that four black freshmen from North Carolina's A & T College had sat-in at the white lunch counter of Greensboro's F. W. Woolworth department store and demanded service hit me like a shot of adrenalin. I was elated. That "something's-gotta-give" feeling that had been building up for years had finally burst forth. The four freshmen, Ezell Blair, Jr., David Richmond, Franklin McCain and Joseph McNeill, were kindred souls.

When reporters asked them, "How long have you been planning this?" they replied, "All our lives!"

I knew exactly what they meant.

Although Gwendolyn was excited about the sit-in, it didn't affect us the same way. She simply wanted to see what would happen; I had a burning desire to get involved. Our heads were just in different places. My identification with the demonstrating students was immediately personal.

Each evening when it was time for the news, I would rush

over to Voorhees, where I was enrolled in high school. I liked to watch the broadcasts in the lounge of the student union. I was always moved by the aura of tension and concern that enveloped the scores of students when the news announcer began to describe events in Greensboro. With the exception of the announcer's voice, the lounge would be so quiet you could hear a rat pissing on cotton.

Hundreds of thoughts coursed through my head as I stood with my eyes transfixed on the television screen. My identification with the demonstrating students was so thorough that I would flinch every time one of the whites taunted them. On nights when I saw pictures of students being beaten and dragged through the streets by their hair, I would leave the lounge in a rage.

The sit-in tactic provided young blacks with a highly effective political weapon. Prior to its use at Greensboro, we had no effective means to express our discontent with segregation and other forms of racial discrimination. All our efforts to confront the system before the sit-ins were vague and unfocused. When a young black attempted to protest some particularly galling aspect of the system before the sit-ins, he would be arrested and charged with "disorderly conduct" or "disturbing the peace."

Our parents had the NAACP. Its practice of pursuing "test cases" through the courts, using laws and the Constitution to fight racial discrimination was suited to their temperaments. We needed something more. As far as we were concerned, the NAACP's approach was too slow, too courteous, too deferential and too ineffectual.

In thousands of instances parents, frightened of reprisals, tried to get their children to quit demonstrating. For the most part, they were unsuccessful. The Negro college students leading the demonstrations and the thousands of nonstudents assisting them were responding to something that transcended parent-child relationships.

John Lewis, a young seminary student at Fisk University in Nashville, whom I later got to know and love like a brother, sent a terse note to his parents when they demanded that he quit demonstrating: "I have acted according to my convictions and according to my Christian conscience. . . . My soul will not be satisfied until freedom, justice and fair play become a reality for all people."

John's note was typical. Young blacks across the South were addressing their parents in the same "this-is-something-I-am-going-to-do" manner. Most parents got the message.

Denmark had its first sit-in about two weeks after the movement began in Greensboro, which was about 225 miles to the north. It was led by Churchill Graham, president of Student Governing Association at Voorhees. Accompanied by six or seven students, Churchill sat-in at a small drugstore in downtown Denmark. The drugstore was picked because it had a segregated lunch counter and because students at Voorhees and SCAT spent a lot of money there.

Although I was at least three years younger than Churchill and the other students who participated in the sit-in, I was in on the planning sessions for it. They permitted me to join the sessions because they liked me and because I was thoroughly familiar with downtown Denmark.

I had a long talk with my mother the night before the demonstration. I knew enough about the attitudes of Denmark's whites to suspect that they would try to put pressure on the parents of those students involved in the demonstration. Although I wasn't going to be sitting-in at the drugstore, I knew that my name would probably be mentioned during the subsequent investigation. I thought it would be best if my mother heard about my involvement from me instead of the local newspaper.

I told her just before she went to bed. I was very nervous because I was not sure of what her reaction would be. I was aware that she had been keeping up with the demonstrations in Greensboro and that she understood the dangers involved in our planned demonstration. She didn't say anything while I told her of our plans.

When I finished talking, I looked deep into her eyes seeking approval. She smiled tenderly and told me to be very careful. She also told me not to worry about pressures being put on her and Daddy. I went to bed that night sure that the demonstration was going to work out fine.

Given all the planning we put into it, the demonstration was anticlimactic. The students marched into the drugstore, walked quietly to the lunch counter, took seats and tried to order food. The world did not explode. The startled whites seated at the other end of the lunch counter did not pull out double-barreled shotguns and begin to assassinate them. Everyone was very calm.

The waitress walked over and told them that they were seated at the "white lunch counter" and that Negroes were not served there. When they continued to demand service, she went and got the manager. When they refused to leave, he called the police.

Denmark's finest were befuddled. They were nervous and unsure of themselves. After conferring with the manager of the drugstore, they walked slowly over to the lunch counter and placed the demonstrators under arrest. The students, who were prepared for every contingency, including arrest, remained calm.

The arrests precipitated an uproar. Everyone on the campus wanted to know what had happened. Students from SCAT flooded the Voorhees campus trying to get news. The Voorhees

Alumni Association was furious. One of the association's representatives called the police and demanded to know why the students had been arrested and how much money it would take to get them released. When he was informed that they hadn't been charged with anything, he hit the ceiling. The entire black community began to mobilize in support of the students.

The police were further confused by the reaction to the student arrests. After a while, they decided to release them. They called Voorhees and Father Grant went down to the station and got the students.

White Denmark was dumbfounded. Blacks had never confronted the system in such a united manner before. A call was sent out for additional law-enforcement officers. By nightfall, the city was swarming with hundreds of state troopers and national guardsmen.

We had a big meeting at Voorhees the next day. Churchill and the guys who had sat-in with him at the drugstore were the principal speakers. Just before the meeting adjourned, Churchill asked for volunteers to accompany him on another demonstration later in the day. Seventy-five students stood and vowed their support.

All those students who intended to participate in the demonstration got dressed up. Most of the boys wore white shirts, suits and ties. The girls wore stockings and heels; some wore gloves.

We were very concerned about image. We wanted to show white people that we knew how to act and were, therefore, worthy of eating where they ate. We were so conscious of image, in fact, that we probably would have turned down anyone who attempted to join the demonstration while attired in a sweat shirt and dungarees.

The demonstrators had intended to take the bus that Voorhees operated on the town-to-campus run, but they got a rude surprise;

the Board of Trustees canceled the bus. The trustees, who did not want the students to conduct the demonstration, mistakenly thought that it would be thwarted.

The students lined up in twos and marched the three dust-filled miles to town. There were hundreds of black people standing along the march route. Some yelled encouragement, but most stood in awed silence. The marching students were quiet too. It was very dramatic.

The city police, accompanied by state troopers, national guardsmen and a large segment of the town's white populace, were stationed along the march route. You could tell by the look in their eyes that they were frightened. Nothing in their past experience had prepared them for such an occurrence.

When the students reached the drugstore, they got their biggest surprise of the day. It was closed—the proprietor had received word that they were on their way. After conducting a quick strategy session in front of the locked doors of the drugstore, the students reformed their lines and marched back to the campus.

Back on campus, they were met by the president of Voorhees, John Potts. His words came as no surprise. He told us that the Board of Trustees had conducted an emergency meeting and decided that there shouldn't be any more demonstrations.

Like every man who was president of a Negro college at the time, President Potts was on the spot. The Board of Trustees, which included people who regularly made sizable financial contributions to the school, were telling him to put a halt to the demonstrations. At the same time, he was faced with a potentially rebellious student body bent on exploiting every opportunity to continue the newborn movement.

We were not particularly troubled by President Potts's personal dilemma. As far as we were concerned, he was faced with a moral decision. We were on the right side of a gigantic moral chasm created by greed, prejudice and ignorance. The Board of Trustees was attempting to act as if a middle ground existed. We knew that there was no middle ground. We also knew what we had to do. President Potts had to decide what he was going to do.

We were faced with a tactical decision. Some students wanted to ignore President Potts and launch a series of attacks on segregation in downtown Denmark. Others felt that it would be improper to try to conduct downtown demonstrations before we made things right at Voorhees. We decided, after much discussion, to concentrate on the school.

A list of demands was drawn up and submitted to President Potts. We wanted better food in the cafeteria, an end to compulsory class and chapel attendance, later curfews, an end to compulsory dress regulations and the right to demonstrate against off-campus racial discrimination when we saw fit.

We spent the remainder of the school year locked in struggle with the Voorhees administration. President Potts and the Board of Trustees refused to concede to our demands. They thought that we would lose interest over the summer, but they were wrong. When we returned to school in September, we took up right where we left off. In mid-October, the Alumni Association stepped in and arranged a settlement. In return for administrative concessions on most of our demands, we agreed to end all campus demonstrations.

By November, just before my sixteenth birthday, I emerged as the leader of the Voorhees student movement. Since we didn't have a formal organization, my position was unofficial. Nevertheless, students on the campus looked to me for direction. I don't know exactly why; it was probably a combination of things. I was well known, popular, a member of the basketball team, an honor student and zealously committed to the movement.

My most immediate concern was to redirect the emphasis of the Voorhees movement from the campus to downtown Denmark. This did not prove to be an easy task. Denmark's whites, who had immediately perceived the symbolic significance of the drugstore where the first demonstration was held, had removed all the seats from the white-only lunch counter. Anyone who wanted to stand up and eat was welcome to do so.

Although some students considered the removal of the seats a victory for us, most of us knew better. Denmark was still very much segregated. While trying to pick a suitable target for demonstrations, I received an invitation to visit a group of students who had been jailed for demonstrating in Rock Hill, North Carolina.

The people I met in Rock Hill, which is 150 miles north of Denmark, were inspiring. They had been protesting for months and had endured many hardships. Local whites, who had attacked demonstrators with ammonia bombs, were determined to break the back of the movement. Rock Hill's blacks were just as determined to keep going. One of their leaders, the Reverend C. A. Ivory, exemplified their fierce commitment.

Reverend Ivory, who was not young anymore, was crippled and could not get about without a wheelchair. Nevertheless, he was on the picket lines every day. When whites called his home in the middle of the night and threatened to kill him if he did not stop protesting, he ignored them. One day he was arrested while sitting-in at a segregated lunch counter. The police wheeled him out of the store and across the street to the jail.

After being searched and fingerprinted, Reverend Ivory was wheeled into a cell and locked up. He remained there for several hours until his attorney bailed him out. The next day he was back on the picket line.

One of the student demonstrators was named Ruby Doris Smith. She was a member of a new organization called the Student Nonviolent Coordinating Committee (SNCC). Ruby Doris was medium height with plain features. She had been a sophomore at Spelman College in Atlanta when the Freedom Rides began.

I was tremendously impressed with her descriptions of the work that SNCC was doing with college students across the South. Little did I know that I would in time be closer to her than I was to my own sister.

I left Rock Hill with the realization that our chances for launching a successful assault on segregation in Denmark would be greatly enhanced if we could attract community support. Rock Hill's blacks did everything they could to aid the demonstrating students. They opened their churches for student meetings, lent their cars and provided students with all the food they could eat. Ruby Doris gained eighteen pounds while serving a thirty-day sentence as a result of eating the rich foods the community people brought each day.

After a few strategy sessions, I decided that the best way to win similar support from Denmark's blacks was to coordinate our efforts with the only established civil rights organization in town: the NAACP. When we suggested to the local NAACP officials that we wanted to work with them, they said that questions of segregation and other forms of racial discrimination should be settled in the courts and not in the streets. They argued that marches and demonstrations were too provocative and, in the final analysis, counterproductive.

"You young people are trying to go too fast," they told us. "If you'll just be patient, the courts will decide these matters."

Afterward, I thought about the familiar glint I'd seen in the eyes of Denmark's reluctant NAACP officials. I readily recognized it because I'd seen it in the eyes of blacks all my life. It was fear. They were afraid to work with us. It was as simple as that.

I was stuck for a while. But after considering and discarding a number of proposals, I came up with a plan. I would bypass the NAACP's local and state officials and apply for a youth chapter charter from the organization's national headquarters in New York. Father Henry Grant from Voorhees agreed to sponsor the application. We didn't tell anyone about it until we got the charter. Little did I know when I agreed to be chairman of the chapter that my involvement would virtually destroy the relationship between me and my father for a long time to come.

BECAUSE I WAS chairman of the youth chapter, a great deal of attention was lavished on me. Most of that attention was critical. Father Grant stood behind me, however. Even when it became obvious that the administrators of Voorhees, his employers, strongly disapproved of me and the youth chapter, he refused to temper his support.

I became aware during that period that subtle pressures were being placed on my mother at SCAT. A number of the people she worked with were telling her that I was going to "destroy" everything my father was working so hard to build for me and Gwendolyn.

Although she told me about the things people were saying to her at SCAT, my mother did not try to get me to stop organizing the youth chapter. When I asked her if she wanted me to stop, she told me to continue doing what I thought was right.

I knew that the shit was about to hit the fan when I heard that Voorhees had fired Father Grant. I realized that we had to do something fast. Why not hold a big rally with a big out-of-town speaker? We could demonstrate to Denmark's older blacks that we had outside support and we could focus national attention on local problems.

I discussed my plan with Joyce Jeeter, one of the Voorhees students who had been active in the student movement from the very beginning. She was enthusiastic. "It sounds like a good idea to me. In any event," she added, "we don't have anything to lose."

It took us almost a month, but we got it all together. The rally was to be held on a Sunday afternoon in one of the biggest black churches in Denmark. We put signs up all over the black side of town so that everyone would know the exact time and place.

The speaker was from the NAACP's regional office in Atlanta. Her name was Julie Wright and she was a lawyer. It wasn't every day that a "big-time lawyer from Atlanta" spoke in Denmark; people were excited.

Joyce and I felt especially good. Our work was paying off. We knew that we were on the verge of pulling off one of the most important political meetings in Denmark's history. I went to sleep the night before the rally filled with excitement. I was also a little nervous. I had never spoken before a large group of people and was concerned about getting stage fright and forgetting my carefully prepared speech.

I woke up early. After hurriedly putting on my clothes, I rushed outside to check the weather. It was perfect. The sun, which had been up about an hour, was burning the last few drops of dew from the grass. The cloudless sky was a clear, robin's-egg blue. It's going to be a beautiful day, I said to myself before going back into the house.

After breakfast, I went into the living room and started to go over my speech and my checklist for the last time. Miss Wright was to come to the house first and I was to escort her to the church. I wanted everything to be in order when she arrived. I was halfway through my speech when my father came into the room.

I knew from the sound he made in the back of his throat just before speaking that something was wrong. He did not waste any time telling me what it was. "I want you to stop what you've been doing," he said.

I was incredulous. Although I knew exactly what he meant, I asked, "What do you mean?"

"I mean I want you to stop. I want you to stop everything you've been doing. Right now. I think you've gone as far as you should go."

I still couldn't believe that he would do this to me. Stop? I couldn't stop. I had committed myself. My reputation was on the line. All my friends—the whole town—knew about the rally, about the youth chapter. I was the chairman.

"Why?" I asked. "Everybody is depending on me. I can't stop."

"Yes, you can," he replied. "I've been working all my life to build something for you. This demonstrating and rallying is no good. If you keep on, you're going to destroy everything. Let some of the others do it. You've done your part."

I was angry. He had no right. I wanted to say, "Goddammit, nigger, you're scared! You're scared of what those candy-assed white crackers will do to you!"

That's what I wanted to say. Instead, I asked him another question.

"What do you suggest I do?"

I was hoping that he would empathize with my dilemma and allow me to attend the rally and bow out of the movement afterward.

"I suggest that you stay home and not attend that meeting this afternoon," he said. "I want you to stop everything right now!"

While he was talking, one of his friends came into the house. The man told me that my father was right, that the youth chapter was a bad thing that would almost certainly cause trouble for my family and me. The man worked at SCAT and he had heard of some of the problems my mother had been having because of my activities.

It was all I could do to keep my wits about me. Wow, I thought, these guys are trying to freak me out! My mother and my sister were there. They heard everything that was being said, but there was nothing they could do. My father's word was law.

Miss Wright arrived at 2 P.M., an hour before the rally was to begin. It was terrible. Everybody but my father was extremely upset. We just sat around as though we were made of wood. I tried to introduce her, but I didn't know how to handle the situation. I was very awkward and that made things worse. It was very quiet after the introductions. Nobody knew what to talk about. My father sat staring alternately at me and at Miss Wright. He wasn't discourteous, just cold.

Joyce arrived after a while. I had arranged for her to take my place as master of ceremonies at the rally. She was upset too. She had not expected to have to speak at the rally. My incoherent instructions probably added to her confusion. As she left the house with Miss Wright, I decided to try one more time to get my father to change his mind.

He was adamant.

"Well, what am I supposed to be doing?" I asked in desperation.

"Nothing," he said.

I received reports on the rally afterward. Joyce handled things well and Miss Wright gave a good speech. The youth chapter never did get going, however. Joyce was not able to do everything and I was forbidden to assist her. Moreover, most of Denmark's black parents were handing down ultimatums similar to my father's. All of them were frightened.

I was ready to leave home, but I didn't know where to go. Every time I saw my father, I was reminded of what he had made me do. I couldn't forgive him. I could see no logical reason why I should remain in his house any longer. I started looking for a boarding school. After spending a couple of months searching through brochures in the Voorhees library, I found a school that seemed to offer everything I wanted.

Although my parents were not enthusiastic about my leaving home, they said that they would honor my wishes. When the application arrived, I immediately filled it out and sent it back.

I didn't see any reason why the school would not accept me. I had excellent grades, Voorhees had an excellent academic rating and I had excellent recommendations from my instructors. It's just a matter of time. You'll be gone in a few days, I told myself.

I was sitting at home one November afternoon when I got a long-distance telephone call. My hand was shaking when I took the receiver.

"Mr. Cleveland Sellers?" asked a formal voice on the other end of the line.

"Yes, I am Cleveland Sellers," I replied.

"We have your application and everything looks very good," said the voice.

I began to smile.

"One thing puzzles us, however," the voice continued. "You left the race space blank. Are you colored?"

My smile vanished. When I answered, "Yes," the voice got a bit brusque.

"I'm sorry to inform you, Mr. Sellers, that we do not have any more openings."

That was it. I was stuck. It was too late to apply to another school. For another year I was going to have to remain in Denmark, at Voorhees, near my father.

When I graduated from high school, I had two goals: a college education and active involvement in the movement. My counselors at Voorhees urged me to apply to a white college in the North. They told me that my chances for admission were excellent. "Your grades are good enough to get you admitted and your parents can afford to pay the high fees. Furthermore," my counselors added, "those white schools up North are looking for *qualified Negroes* like you. They'll probably give you a full-tuition scholarship!"

Despite their attempts to persuade me, I had absolutely no interest in Northern white colleges. "I am a black Southerner," I said, "I am going to attend a black college in the South. I want to remain with my people, where the action is. I intend to be a part of the movement. I can't do that while going to school at some white college up North."

After studying a number of Southern black colleges, I decided that Howard University in Washington, D.C., was the school for me. It had a good academic program, and, more important, its students appeared to be very active in the movement. My mother was upset when I announced that I was going to attend Howard. She wanted me to attend South Carolina State because it was closer to home, but she went along with my decision. My father said that he didn't care what college I attended as long as I studied hard, made good grades and did not waste his money.

CHAPTER 3

SNCC: The Beginning

WITHIN THREE MONTHS after its electric beginning in Greensboro, the sit-in movement was so large that it was virtually impossible to keep tabs on everything happening. Students were conducting kneel-ins, worship-ins, wade-ins and sit-ins in more than 125 Southern cities. Arrests and jailings ran to the thousands. In most cases, the only communications between different groups of demonstrating students were radio, television and newspaper accounts. The need for some form of coordination was clearly apparent.

Ella Baker, the first executive secretary for Dr. Martin Luther King's Southern Christian Leadership Conference (SCLC), made the most important response to that need. She acquired eight hundred dollars from SCLC to sponsor an organizing conference for the demonstrating students.

The conference was held April 15–17, 1960, on the campus of Shaw University in Raleigh, North Carolina, Miss Baker's alma mater. It was the weekend of spring vacation, a period that Negro college students traditionally reserved for boisterous parties and fund-raising expeditions to their parents' homes. Now, however, it was a time for breaking with established traditions. Over three hundred students attended the conference.

"Two hundred more than we expected," recalled Miss Baker.

I was still in Denmark attending Voorhees High School when the Raleigh conference was held. My most immediate concern at the time was helping organize students to confront the Voorhees administration. Because this book combines my biography and the history of SNCC, I have had to reconstruct the events that took place at the conference and during the period covered in this and the following chapter. I used books, magazine and newspaper accounts, interviews* and the memories of early movement veterans whom I got to know in later years.

At the point of death, people are sometimes asked what they would do differently if they could live their lives again. If asked that question when my time comes, I am certain that one of the things I will say is, "I would like to have been three years older in April, 1960, and been a delegate at the Raleigh conference."

The delegates at the conference represented fifty-eight Negro colleges from twelve states. Nineteen delegates from white Northern colleges were also in attendance. Only twenty-five of the three hundred delegates were personally acquainted.

"The rest of us knew each other only by reputation," recalled Julian Bond, who was then representing an Atlanta student group called the Committee on Appeal for Human Rights. "You knew that the Nashville group was the first one out, even though the Greensboro people sat-in first. The Nashville people had been in

* I am particularly indebted to Julian Bond. In an interview with me and Robert Terrell on September 20, 1970, he provided much valuable information on the events that led to the formation of SNCC and its initial development.

vigorous training, undergoing vigorous discipline. You knew that the Atlanta group was wealthy and cocky; you knew that the Alabama group from Montgomery had been bruised the most; they had been brutalized. Everybody had his own little rep."

The conference marked the beginning of a new epoch. The student delegates, filled with pride and excitement, were acutely aware of the unparalleled sense of brotherhood that existed among them. They were united by a common goal against a common foe. In later years, Jane Stembridge, one of the few whites in attendance, described the mood of the three-day gathering. "It was," she said, "the purest moment."

The three most important speakers at the conference were Dr. Martin Luther King, Jr., James Lawson, Jr., leader of the Nashville students, and Miss Ella Baker. Dr. King emphasized three things in his speech: the establishment of an ongoing organization to lead the struggle, a nationwide campaign of selective buying to reward progressive business establishments and punish those that remained segregated and the formation of an army of volunteers prepared to accept prison sentences in lieu of fines and bail.

Dr. King also told the students that nonviolence was not a tactic, but a way of life.

"Our ultimate end must be the creation of the beloved community. The tactic of nonviolence without the spirit of nonviolence may become a new kind of violence."

Although he was extremely popular with the students, Dr. King's address was not as well received as James Lawson's. Lawson, who had recently been expelled from Vanderbilt University in Nashville, Tennessee, for his role in the Nashville student movement, was a great hero to the students. Some of them referred to him as "the young people's Martin Luther King."

Like Dr. King, Lawson was an advocate of nonviolence. There were some important differences between them, however. Lawson was not as troubled as Dr. King by those who embraced nonviolence for tactical reasons. Although he was a minister, he did not insist as did Dr. King that those who took part in the movement be "disciples of nonviolence."

Lawson's speech, which was entitled "We Are Trying to Raise the Moral Issue," reflected the impatience of his youthful listeners.

"The nonviolent movement is asserting 'get moving,'" charged Lawson. "The pace of social change is too slow. At this rate, it will be at least another generation before the major forms of segregation disappear. All of Africa will be free before the American Negro attains first-class citizenship. Most of us will be grandparents before we can live normal lives."

Lawson received a standing ovation when he finished. The cheering, clapping students were in complete accord with his insistent militance.

Miss Baker's speech received a different response.

"Although it was not well received, her speech was probably the best of the three," said Julian Bond. "Her thesis was 'More Than a Hamburger'; that what we were up to was more than just eating a hamburger at a lunch counter. She said that we ought to be interested in a whole range of social problems that dealt with black people; that lunchcounter protests were the first step towards involving ourselves in this whole process. She told us that there were some good and some bad old people, but on the whole, they would try to co-opt and destroy what was basically our movement; a young people's movement.

"We largely ignored her advice because we said, 'Aw, ain't nothing more to it than a hamburger. If we can eat this hamburger, everything will be straight. Everything else will fall into place.'"

There was a great deal of pressure on the students at the conference to affiliate with one of the older civil rights organizations. SCLC, CORE and the NAACP were well represented.

CORE, which was the smallest and least known of the older organizations, had no base in the South. Its representatives tried to get the students to turn their separate campus organizations into Southern CORE chapters. If the students had accepted the plan, CORE would have been transformed into the most important civil rights organization in the nation.

The NAACP, which was initially critical of the sit-ins, had begun to take credit for them. Many of the demonstrating students belonged to NAACP youth chapters and the organization's leaders saw the sit-in movement as a fund-raising catalyst. The students remained unimpressed with the NAACP, however, and rejected it.

SCLC, which had been stagnating since the 1956 Montgomery bus boycott, needed the students badly. Dr. King and his associates realized that the students had what SCLC did not have: a program. Moreover, the nonviolent character of the student movement was consistent with SCLC's philosophy. After several heated discussions, the students decided to reject SCLC's overtures also.

Julian Bond explained the reasons. "We resisted the affiliation requests of the older organizations partly because of Miss Baker's feeling that we didn't need to become a part of an existing organization and partly because it was very heady stuff for young people seventeen and eighteen years old to be running their own political organization.

"You were running your own little group. You had your own office. You may have had your own bank account. *You* made decisions. You sat down with whoever was the biggest nigger in town before you came along. You spoke with white folks, made them tremble with fear. It was very heady stuff."

Before adjourning the conference, the students established a Temporary Coordinating Committee charged with meeting monthly through spring and summer and coordinating communications between campuses where demonstrations were occurring. Ed King, a student from Kentucky State in Frankfort, was asked to serve as the committee's administrative secretary.

"Ed King had a bloody shirt," explained Julian. "He had been beaten quite badly at Kentucky State. It sounds funny now, but a lot of people used to do that; carry those shirts around with them. He didn't wear it, but had it with him, never been cleaned.

"It was like Jackie Kennedy's wearing the dress with her husband's blood on it. Her theory was, 'I want them to see. I want them to see his blood.' Our theory was, 'We want them to see *our* blood. You know, this is *my* blood on this shirt!'"

In May, 1960, fifteen of the students who attended the Raleigh conference held a second meeting at Mount Moriah Baptist Church near the campuses of the Atlanta University complex. The meeting was also attended by Dr. Martin Luther King, Miss Ella Baker, James Lawson and representatives from several white liberal organizations headquartered in Atlanta. A number of important decisions were made at the meeting.

The students decided to set up an office, hire a secretary and change the name of the Temporary Coordinating Committee to the Temporary Student Nonviolent Coordinating Committee. They also decided to offer testimony to the Democratic and Republican Platform Committees later in the summer. Marion Barry, a graduate student at Fisk University in Nashville, was asked to serve as the new chairman.

Marion Barry's election constituted a reaction against two

things: the Church and Northerners. Barry was a Southerner and he was not a minister. Many students in the organization had begun to resent the quasi-religious orientation of the movement. Although they respected Dr. King, they disliked his tendency to merge his political and religious roles.

Barry's Southern origin was more important to the students than his not being a minister. As early as May, 1960, there was resentment against the aggressive advice of Northerners, especially white Northerners. By picking Barry to head *their* organization, the students were serving notice to Northern whites—the National Student Association in particular—that they intended to keep their organization under the control of Southern black students.

Before adjourning the Mount Moriah conference the students adopted a one-page statement of purpose:

> We affirm the philosophical or religious ideal of nonviolence as the foundation of our purpose, the presupposition of our faith, and the manner of our action. Nonviolence as it grows from the Judaeo-Christian tradition seeks a social order of justice permeated by love. Integration of human endeavor represents the crucial first step towards such a society.
>
> Through nonviolence, courage displaces fear; love transforms hate. Acceptance dissipates prejudice; hope ends dispair. Peace dominates war; faith reconciles doubt. Mutual regard cancels enmity. Justice for all overcomes injustice. The redemptive community supersedes systems of gross social immorality.
>
> Love is the central motif of nonviolence. Love is the force by which God binds man to himself and man to

> man. Such love goes to the extreme; it remains loving and forgiving even in the midst of hostility. It matches the capacity of evil to inflict suffering with an even more enduring capacity to absorb evil, all the while persisting in love.
>
> By appealing to the conscience and standing on the moral nature of human existence, nonviolence nurtures the atmosphere in which reconciliation and justice become actual possibilities.

The statement of purpose, which was drafted by James Lawson, was distributed across the South with the following message: "We urge all local, state or regional groups to examine it closely. Each member of our movement must work diligently to understand the depths of nonviolence."

In July, 1960, Marion Barry and three other students traveled to Los Angeles to speak before the Democratic National Convention's Platform Committee. They took a train because they could not afford to fly.

The students decided to appear before the Platform Committee for three reasons: (1) the sit-ins were a hot political issue, (2) people around the nation were trying to decide if the sit-ins were legitimate political protests or anarchy, and (3) the Democratic party seemed to be interested in doing something constructive about civil rights. The goal of Marion Barry and his three associates, therefore, was to let the nation know that if *any* decisions were made about the sit-in movement, the Temporary Student Nonviolent Coordinating Committee *had* to be taken into consideration.

"We wanted them to know that *we* were the people who were causing all the trouble, that *we* were the ones who were sitting-in and getting arrested, that *we* were the ones who were running the boycotts," Julian Bond said.

Barry's comments before the Platform Committee reflected the combination of indignation and "poetic freshness" that characterized that stage of the student movement.

". . . the ache of every man to touch his potential is the throb that beats out the truth of the American Declaration of Independence and the Constitution," he declared. "America was founded because men were seeking that room. . . . We want to walk into the sun through the front door. For three hundred and fifty years, the American Negro has been sent to the back door."

Some critics of the student movement, including former President Harry Truman, had accused the students of being Communist-inspired. Barry responded to their charges. "To label our goals, methods and presuppositions 'Communistic' is to credit Communism with an attempt to remove tyranny and to create an atmosphere where genuine communication can occur. Communism seeks power, ignores people and thrives on social conflict.

"We seek a community in which man can realize the full meaning of the self, which demands open relationship with others."

The Temporary Student Nonviolent Coordinating Committee got its first full-time, paid staff member in June, 1960. Her name was Jane Stembridge. She was a white Southerner just graduated from Union Theological Seminary in New York. Her office was a cluttered corner in SCLC's headquarters in Atlanta.

Jane spent most of her first month getting out mailings to Negro college campuses across the South. In late July she was joined by Robert (Bob) Moses. Bob, who had a master's degree in philosophy from Harvard University, came South looking for a nonexistent job with SCLC.

"This quiet soul walked into the office and explained that someone in New York had told him there was a job with SCLC waiting for him in Atlanta," said Miss Baker. "But there had been some foul-up in communication, so we put Bob to work licking envelopes for the student group. He and Jane hit it off immediately, both of them being philosophy students, and they talked about Camus, Tillich and Kant all day."

Although he agreed to work for almost no salary, Bob Moses was not readily accepted by the leadership of the Committee.

"We were immensely suspicious of him," Julian recalled. "All this is very embarrassing to relate years later, but here is this guy much more clever than we. He had a much broader view of social problems and social concerns than we did. We had tunnel vision. We looked straight ahead. 'Lunch counters here—that's a problem. Employment over here—that's a problem.' We were convinced that we could knock them over with little effort and everything would then be okay.

"Bob Moses, on the other hand, had already begun to project a systematic analysis; not just of the South, but of the country, the world. He didn't try to impress it on us. He didn't say, 'Here's what's right, you've been doing this wrong.'

"Rather than think there was something wrong with us, we thought there was something wrong with him. We just couldn't figure why a guy from New York would come down here."

Bob Moses believed that the student movement should leave the urban centers of the South and move into rural areas where no one else was working. Atlanta, for instance, was the headquarters, either regional or national, of every major civil rights organization in the nation. Moses reasoned, therefore, that the students should leave Atlanta. He felt that leaving would accomplish two things:

the elimination of the constant conflicts with other civil rights organizations working in Atlanta and assistance for rural blacks who were not being helped by anyone.

The Temporary Student Nonviolent Coordinating Committee held a big conference in Atlanta in October, 1960. The conference was sponsored by grants from Northern students, SCLC and the Packinghouse Workers Union. The Packinghouse Workers added a stipulation to their grant when they discovered that Bayard Rustin, a radical black who had been previously accused of being a Communist, was scheduled to speak at the conference. They told the students that they considered Rustin an inappropriate speaker. They also told the students that their grant would not be forthcoming unless Bayard Rustin's invitation was withdrawn.

After a series of hasty meetings, the students decided that they needed the Packinghouse Workers' grant more than they needed to hear Bayard Rustin. A telegram was sent to Rustin asking him not to attend the conference.

"It was a decision which many of the participants are embarrassed about now," recalled Julian Bond, who was destined to become the central figure in a major freedom-of-speech controversy with the Georgia legislature in later years.

"You have to figure who we were," Julian added. "We were people with no concept of civil liberties. We really were much into the hamburger. That was the issue; the hamburger. We said, 'Well, if Bayard Rustin is objectionable to people who are going to give us money, he's got to go!'"

Jane Stembridge resigned from SNCC over the Rustin issue. She was outraged by the decision to ask him not to attend the conference. Plans for the conference went on despite her resignation, and telegrammed invitations subsequently were sent to the

Democratic and Republican presidential candidates, Richard M. Nixon and John F. Kennedy:

> *The Student Nonviolent Coordinating Committee invites your participation in its second annual general conference to be held October 7–9 in Atlanta. This conference will be attended by some 300 negro student sit-in leaders, Northern student support groups and representatives from local and national human relations organizations. It will provide the moral impetus for the renewal of nationwide nonviolent student protests against second class citizenship. You are invited to meet the students who must continue these protests until definitive civil rights action is taken.*
>
> *Sincerely yours,*
> *The Student Nonviolent*
> *Coordinating Committee*

Additional wires were sent to vice-presidential candidates Lyndon B. Johnson and Henry Cabot Lodge, Republican National Chairman Thruston B. Morton and Democratic National Chairman Henry Jackson. The wires stressed the "imperative need for immediate legislative action to end racial segregation and discrimination in all areas of public accommodations."

The two most important things that took place at the October conference were the dropping of the word *Temporary* from the name of the organization—the Student Nonviolent Coordinating Committee (SNCC) was officially born—and Charles McDew's election to the chairman's post. McDew was elected because Marion Barry resigned in order to return to school. The word

Temporary was dropped because the students realized that they were in for a long, hard struggle.

McDew, who had been active in the student movement in Orangeburg, South Carolina, was a natural leader and a very good speaker. Less than one month after he was elected, he delivered a major address at the Antioch College Conference on Human Rights. His speech reflected SNCC's growing political sophistication and increased militance.

"Our fight," declared McDew, "is not against persons, but persons involved in the promotion and perpetuation of the system we would revise. The sit-ins have inspired us to build a new image of ourselves in our own minds. And, instead of sitting idly by, taking the leavings of a sick and decadent society, we have seized the initiative, and already the walls have begun to crumble."

In conclusion, McDew promised his listeners that the Southern student movement would not end until "every vestige of racial segregation and discrimination are erased from the face of the earth."

The new self-image to which McDew referred was very real. For by October, 1960, SNCC's members realized that they were more than a group of "temporary" protesters and demonstrators. They were, instead, the militant leaders of a new movement.

Theirs was the politics of direct action. They were not guided by grand political theories, however, only by their tenacious belief in the moral rightness of their cause. Their unofficial motto was Do What the Spirit Say Do! Their official motto: We Shall Overcome!

When a Black Muslim minister suggested that they develop a greater sense of racial pride, they were amused. He obviously doesn't understand *our* movement, they thought. When the minister advanced a "weak Marxist" analysis of the South and racial discrimination, they remained unimpressed.

"It wasn't that there was no one to our left; there was," Julian Bond recalled. "We just were not ready. We were into our own thing. We *knew* it was right. To our mind, lunch-counter segregation was the greatest evil facing black people in the country and if we could eliminate it, we would be like gods."

CHAPTER 4

Courage, Commitment and Major Responsibility

BEFORE THE FREEDOM Rides, which were initiated in the spring of 1961, SNCC was one of many small and relatively anonymous black organizations struggling to make a contribution to the cause of black liberation. Although it was closer to the burgeoning student movement than any other civil rights organization, many people still believed that SCLC and the NAACP were coordinating the student demonstrations. After the Freedom Rides, there was little doubt in anyone's mind about who was leading the student movement.

The reason SNCC became a major civil rights organization after the Freedom Rides is rather simple: it was the only action-oriented civil rights organization in the South prepared to absorb the brash young militants who joined the movement because of the Rides.

Although CORE initiated the Rides, it was not prepared to absorb new members in the South. It still didn't have a Southern base. The NAACP was not interested in recruiting the militant youngsters who, for the most part, disdained its legalistic approach to the struggle. SCLC, with its moderate approach to protest, was not enthusiastic about recruiting the volatile young militants either.

SNCC welcomed the militant youngsters with open arms. Stokely Carmichael, a philosophy student at Howard University, and John Lewis, a religion major at Fisk University, were among those who joined SNCC during this period.

The students demanded that James Forman be hired as SNCC's executive director. James Forman, a large man with the stature and disposition of a bear, was a former Chicago schoolteacher. Several SNCC members had met him in Fayette, Tennessee, where he was working with a group of poor blacks. Impressed by his mind and organizational ability, the students claimed that they would feel more secure if they knew that a stable figure like Forman was coordinating the "National Office" in Atlanta.

When Forman was offered the job, he agreed to take it—for one year. Although he was thirty-three years old, at least eleven years older than SNCC's average member, he had an immediate effect on the organization. He was totally committed and demanded that the students working with him be the same. He arrived at the office early in the morning and left late at night. Without being asked or told, the students began to do likewise.

He was an organizer, an idea man. His mind seemed always to be occupied with movement strategy, fund-raising schemes, publicity and the safety of the Mississippi-bound organizers. The movement was not a job to Jim Forman: it was his life.

In August, 1961, a small band of organizers led by Bob Moses set out to conduct voter registration drives in rural Mississippi. Bob had quit his job and joined SNCC. Despite Jim Forman's reassuring presence in Atlanta, the organizers realized that they were basically on their own. They were paid only twenty dollars a week each—minus taxes—not enough to buy food, clothing, shelter *and* supplies. If they survived, it would be because they managed to recruit assistance from poor blacks in the communities where they worked.

The venture began badly. The organizers began work in McComb, Mississippi, which is located in an area some observers have compared to the "ninth circle of hell." Although they were welcomed by the local chairman of the NAACP, most of the town's blacks were afraid to have anything to do with them. They knew that McComb's whites were willing to kill in order to keep the organizers from accomplishing their goals.

"For the first two weeks," Bob Moses recalled, "I did nothing but drive around town talking to the business leaders, the ministers, the people in town, asking them if they would support ten students who had come in to work on a voter-registration drive. We got a commitment from them to support students for the month of August and to pay for their room and board and some of their transportation while they were there."

One of the biggest fears that the people had was that the organizers might leave as soon as trouble got started. Many people refused to have anything to do with the SNCC organizers for this reason. Perceiving this, Bob Moses devised a tactic to allay their fears.

"We began in McComb canvassing for about a two-week period," he recalled during an interview several years later. "This means that we went around house-to-house, door-to-door, in the hot sun every day, because the most important thing was to convince the local townspeople that we meant business, that is, that we were serious, that we were responsible."

The canvassing broke the ice. After a couple of weeks, the organizers got a few people to try to register. Because the grapevine functions just as well—and probably better—in rural areas than it does in cities, news of the voter-registration campaign stirred people in nearby counties. Blacks in two adjoining counties, Amite and Walthall, requested that organizers be sent to work with them. This presented SNCC with a dilemma.

Amite and Walthall were two of the most dangerous counties in Mississippi. They had a history of particularly vicious racial oppression. SNCC's organizers wanted to avoid them until they had a better feel for their work and the kinds of responses they could expect from whites. Unfortunately, they felt they could not turn down requests for assistance.

"The problem is that you can't be in the position of turning down the tough areas, because the people would simply lose confidence in you," Bob Moses later explained.

After wrestling with the problem for several days, it was decided that SNCC should send organizers to help anyone and everyone who requested it. This was an extremely dangerous decision that nevertheless had to be made. In mid-August, 1961, John Hardy, one of the students just recruited off the Freedom Rides, was sent into Walthall County to organize a voter-registration campaign.

At about the same time Mississippi's whites initiated a nightmarish reign of terror designed to extinguish the spark of defiance that SNCC's organizers had ignited. On August 15, Bob Moses was arrested by a county officer who accused him of being the man "who's been trying to register our niggers." Bob was taken to court, charged with impeding an officer in the discharge of his duties, fined fifty dollars and jailed for two days. On August 22, after his release, Bob was so severely beaten that eight stitches were required to close the wound in his head. Billy Jack Caston, the sheriff's first cousin, was arrested and tried for assaulting Bob, but the all-white jury acquitted him.

On the twenty-seventh and twenty-ninth of August, five black high school students from McComb were convicted of breach of the peace for sitting-in at a bus terminal and a variety store. They were sentenced to a four-hundred-dollar fine each and eight months in jail. One of the students, a fifteen-year-old girl, was subsequently released, rearrested and sentenced to twelve months in a state school for delinquents.

On September 5, in the town of Liberty, SNCC organizer Travis Britt was beaten by a group of whites on the courthouse lawn. He later reported that one of his attackers hit him more than twenty times. On September 7, the voter registrar in Tylertown pulled a pistol on a SNCC worker and two black men. After being beaten over the head with the pistol, the SNCC worker was arrested and charged with disturbing the peace.

On September 25, Herbert Lee, a black man who'd been active in a local voter-registration campaign, was shot and killed by E. H. Hurst, a white state representative. Hurst was not prosecuted because local officials claimed that he killed Herbert Lee in self-defense.

A McComb judge reflected the attitude of many Mississippi whites toward SNCC and its supporters while sentencing nineteen high school students who'd been charged with breach of the peace and contributing to the delinquency of minors: "Some of you are local residents, some of you are outsiders. Those of you who are local residents are like sheep being led to the slaughter. If you continue to follow the advice of outside agitators, you will be like sheep and be slaughtered."

Not all the violence witnessed by SNCC's organizers was directed at them and their supporters. On September 13, four black men were fishing in the Big Black River near Greenwood when they found the remains of an unidentified black man.* The body was encased in a cloth sack and weighted down with one hundred pounds of rocks. The coroner who examined it judged that it had been in the river for four or five days. He said that it had apparently been thrown off a highway bridge.

Charles McDew wired Attorney General Robert F. Kennedy: "WE URGE THE JUSTICE DEPARTMENT TO IMMEDIATELY INVESTIGATE THE KILLING OF AN UNIDENTIFIED NEGRO MAN IN HOLMES COUNTY. PLEASE SEND FEDERAL TROOPS AND/OR MARSHALS TO HELP RECONSTRUCT MISSISSIPPI BECAUSE THE NAZI-MINDED CITIZENS HAVE DECLARED OPEN SEASON ON NEGROES."

Despite the trumped-up prosecutions, the inordinately long jail sentences, the brutal beatings and the constant threat of death, SNCC's organizers refused to be run out of Mississippi. When they had been so beaten by whites that they were afraid to appear on the streets because their bloody wounds might frighten *the people,* they hid until they could wash and change clothes.

When it was too dangerous to be seen talking with people in the daytime, they waited until night and drove along the rut-filled back roads searching out the roughhewn shanties where *the people* lived. On those nights when cadres of whites were out searching for them with loaded shotguns and pistols, they turned off their headlights and drove by the light of the stars and the moon.

When whites appeared outside their dwellings in the dead of night armed with rifles and torches, they always managed to slip away and find safety elsewhere in the black community—among

* Greenwood is near the stretch in the Tallahatchie River where Emmett Till's bloated corpse was found.

the people. When irate whites burned their offices, they moved on and established new ones. When they could not find new offices because everyone was afraid to rent to them, they lived and worked in tents.

At every moment they lived with tension. It was always there, always stretched like a tight steel wire between the pit of the stomach and the center of the brain. Ulcers, migraine headaches, constipation, nervous twitches; all these maladies plagued them.

When the strain became unbearable, when sanity was stretched to its absolute breaking point—and then a little beyond—they would pile into mud-splattered cars and head for Atlanta. Atlanta was behind the lines, a place where the organizers could unwind for two or three days before returning to the front.

Sometimes when things were going badly and they could not get away, they would pick up a telephone receiver and call someone behind the lines. When they couldn't reach close friends, they called casual acquaintances. They wouldn't want anything in particular, just conversation.

Some organizers found relief by writing letters. They would write old friends, relatives, almost anyone. The letters always understated the problems. They were often filled with wry, sometimes caustic bits of humor. Those who had never been exposed to the vicious racism of rural Mississippi could never, ever, begin to comprehend the taut messages between the lines and words. On November 1, 1961, Bob Moses sent this message to a group of Northern supporters:

> *We are smuggling this note from the drunk tank of the county jail in Magnolia, Mississippi. Twelve of us are here, sprawled out along the concrete bunker; Curtis Hayes, Hollis Watkins, Ike Lewis, and Robert Talbert, four veterans*

of the bunker, are sitting up talking—mostly about girls; Charles McDew ("Tell the story") is curled into the concrete and the wall; Harold Robinson, Stephen Ashley, James Wells, Lee Chester Vick, Leotus Eubanks, and Ivory Diggs lay cramped on the cold bunker; I'm sitting with smuggled pen and paper, thinking a little, writing a little; Myrtis Bennet and Janie Campbell are across the way wedded to a different icy cubicle.

Later on Hollis will lead out with a clear tenor into a freedom song; Talbert and Lewis will supply jokes; and McDew will discourse on the history of the black man and the Jew. McDew—a black by birth, a Jew by choice and a revolutionary by necessity—has taken on the deep hates and deep loves which America, and the world, reserves for those who dare to stand in a strong sun and cast a sharp shadow.

In the words of Judge Brumfield, who sentenced us, we are "cold calculators" who design to disrupt the racial harmony (harmonious since 1619) of McComb into racial strife and rioting; we, he said, are the leaders who are causing young children to be led like sheep to the pen to be slaughtered (in a legal manner). "Robert," he was addressing me, "haven't some of the people of your school been able to go down and register without violence here in Pike County?" I thought to myself that Southerners are most exposed when they boast.

It's mealtime now: we have rice and gravy in a flat pan, dry bread and a "big town cake"; we lack eating and drinking utensils. Water comes from a faucet and goes into a hole.

This is Mississippi, the middle of the iceberg. Hollis is leading off with his tenor, "Michael, row the boat ashore,

Alleluia; Christian brothers don't be slow, Alleluia; Mississippi's next to go, Alleluia." This is a tremor in the middle of the iceberg—from a stone that the builders rejected.

Bob Moses

SNCC's members, most of whom grew up in cities, were deeply affected by the rural blacks with whom they were working. By the winter of 1961, the organization's members had begun to walk, talk and dress like the poor black farmers and sharecroppers of rural Georgia and Mississippi.

One project report listed five rules of "staff decorum." The rules are indicative of the lengths to which SNCC's members were willing to go in order to win respect and support from *the people:* "(1) There will be no consumption of alcoholic beverages. (2) Men will not be housed with women. (3) Romantic attachments on the level of 'girl-boyfriend relations' will not be encouraged within the group. (4) The staff will go to church regularly. (5) The group shall have the power of censure."

Although SNCC paid little attention to the private lives of its members when they came to Atlanta where community mores were lenient, those staff members who violated rules and regulations in "the field" were sternly censured. When an organizer in southwest Georgia got a local teen-ager pregnant, he was given a small sum of money and told to "marry her!"

By the end of 1962, SNCC's members had developed a powerful esprit de corps. They worked, ate, socialized and slept together, read the same books and wore the same kind of clothing. They had a definable lifestyle. They were *SNCC people* and they considered themselves different from those who had not shared their unique experience.

SNCC had become a way of life. Everything in their lives revolved around the organization and the struggle for racial equality. Although they frequently had sharp arguments, especially where tactics were concerned, they were bound together in a community of common commitment.

By the end of 1962 the Hamburger theory—a variation of the federal government's Domino theory—had proved an insufficient focus for that commitment. The desegregation of lunch counters and other public facilities across the South had done nothing to alter the inequitable balance of power between blacks and whites. Although blacks had the right to sit and eat beside whites in most areas, they did not have the power to control the "significant events" that affected their lives.

SNCC's members saw the ballot as a significant power lever. They concluded that if black people had ballot power, they would be in a position to exercise influence in many important areas. After much discussion, they decided to launch a daring program designed to get the vote for poor Southern blacks. Mississippi, "the middle of the iceberg," was chosen as the prime target area.

The program, which was launched in the fall of 1963, called for the organization of a statewide Freedom Ballot. The Freedom Ballot was considered a means of demonstrating the systematic manner in which Mississippi's blacks were being discriminated against.

Two men agreed to become candidates in the unofficial campaign. One of them was Aaron Henry, a forty-one-year-old black pharmacist, army veteran and NAACP leader from Clarksdale, Mississippi. Mr. Henry was the perfect candidate. He was known and respected by blacks throughout the state because of his association with the NAACP.

The other candidate on the Freedom Ballot was Edwin King, chaplain of Tougaloo College. Because Mr. King was one of the few white men in Mississippi who had been active in the movement, he was also well known.

Forty SNCC members worked full time through the summer on the Freedom Ballot. Two weeks before the election, they were joined by forty students from Yale and fifteen students from Stanford. The students were recruited for two reasons: to attract national attention to the Freedom Ballot and to demonstrate to Mississippi's blacks that other people in the country were concerned about their plight.

Mississippi's whites were outraged by the students from Yale and Stanford. To them, the students were "nigger-lovin' outside agitators" bent on destroying their way of life. Several students were beaten by irate whites. And some escaped with their lives only because those who shot at them happened to miss.

Despite white terror tactics, the Freedom Ballot was a huge success. Over eighty thousand of Mississippi's blacks took time out from their daily affairs to vote in the mock election. This was four times the number of blacks officially registered in the state. The huge turnout substantiated what SNCC's organizers, Bob Moses in particular, had been saying all along—that Mississippi's blacks wanted to vote and would vote if they had a chance to do so.

The Freedom Ballot set the stage for one of the most important campaigns of the civil rights movement: the Mississippi Summer Project. The project was also conceived by Bob Moses. He saw it as an offensive that would break the back of systematic white racism in the state of Mississippi. The project was coordinated under the auspices of a new organization: the Council of Federated Organizations (COFO). Although a large percentage of Mississippi's black civic and political organizations were represented in COFO, SNCC was the acknowledged leader.

SNCC's staff worked long hours through the winter and spring of 1963–64 preparing for the Summer Project. The organization had grown to 130 members and all of them were kept busy. By midspring, SNCC drew up a prospectus that was circulated to college campuses across the nation.

It was an urgent call to action, embodying a bold challenge, stated simply and forthrightly:

> It has become evident to the civil rights groups involved in the struggle for freedom in Mississippi that political and social justice cannot be won without the massive aid of the country as a whole, backed by the power and authority of the federal government. Little hope exists that the political leaders of Mississippi will steer even a moderate course in the near future. . . . In fact, the contrary seems true: as the winds of change grow stronger, the threatened political elite of Mississippi becomes more intransigent and fanatical. . . . Negro efforts to win the right to vote cannot succeed against the extensive legal weapons and police powers of local and state officials without a nationwide mobilization of support.
>
> A program is planned for this summer which will involve the massive participation of Americans dedicated to the elimination of racial oppression. . . .

Those of us who read the prospectus, accepted its challenge and went to Mississippi have never been the same.

CHAPTER 5

Howard University: The Beginning of Total Commitment

HOWARD UNIVERSITY WAS a big disappointment. I arrived on campus in September, 1962. Filled with the unbounded enthusiasm peculiar to seventeen-year-olds, I expected to see everyone, students, instructors and administrators, passionately involved in the movement. I was eager to participate in emotion-packed mass meetings, tense strategy sessions and frequent demonstrations. Unfortunately, I didn't find any of these things.

When I attempted to discuss the movement with the guys in my dormitory, they would grunt and change the subject. They were much more interested in cars, fraternities, clothes, parties and girls. They loved to sit in bull sessions for hours discussing them. They also spent a lot of time talking about the high-paying jobs they intended to get after graduating. By the end of my first semester, I felt like an outcast. I tried to discuss my feelings and interests with my instructors—I needed help from someone who understood what I was going through—but they were harder to talk to than my classmates. Their primary concerns seemed to be their cars, their homes, their professional associations and their salaries.

The administrators were no different. They were remote men who never seemed to have enough time really to hear what students were saying. They related to us as if we were cogs in a giant machine: those cogs which did not conform to the machine's program were expendable.

There was a great deal of interest among almost everyone on campus in *the Howard image,* which was designed to create the impression that there were no substantial differences between Howard's students and those at elite white colleges. Students went to absurd lengths to conform to *the image.* The guys wore suits, jackets and ties everywhere—including football games and breakfasts. The girls wore stockings and heels. Many of them refused to date men whose clothes did not fit *the image.*

I refused to conform. I had always been a very casual dresser and saw no reason to change. I liked to wear blue jeans, sweat shirts, army jackets and sneakers. Although I did not relish the outsider-outcast role that was accorded me because of my clothes, I was determined not to change.

"I have the right to dress in any way I please!" I snapped at my roommate one night when he attempted to scold me for wearing blue jeans to a big dance.

"But, Cleve," he responded, "you'll never get a girlfriend. No girl's gonna be caught dead with you if you keep dressing like a refugee from World War I."

My roommate was in love with Howard. He was having a ball. He couldn't understand why I had so much trouble adjusting.

"Fuck it, man," he said to me one night in exasperation after I asked him if he didn't feel some responsibility to try to improve racial conditions.

"Don't confront me with that Martin Luther King shit. Everybody's gotta go for himself and I'm going for me. If niggas down South don't like the way they're being treated, they oughtta leave. I'm not going to join no picket lines and get the shit beat outta me by them crazy-ass Ku Klux-ers!"

"I'm interested in four things," he added. "A degree, a good job, a good woman and a good living. That's all. You and Martin Luther King can take care of the demonstrating and protesting. I have *no* use for them!"

My roommate was typical. Although few stated their feelings so bluntly, most of Howard's students shared his attitude.

I met my first real friend near mid-semester. He was a tall, lanky junior with sparkling eyes and an infectious smile. Although he was from New York, we had many things in common, the most important being our intense interest in the movement. We both had a burning passion to do something about the plight of blacks. His name was Stokely Carmichael.

From the day we met, I considered Stokely a special friend, a special person. Although he was flamboyant and extremely cocky, there was something about his manner that attracted people. Everyone on campus knew him. Few of his admirers ever got close enough to him to see what I saw: an extremely sensitive person who generally disguised his sensitivity with bombast.

Stokely, who had worked in Mississippi for SNCC the previous summer, belonged to a campus organization called the Nonviolent Action Group (NAG). It was just what I had been looking for. I joined immediately.

NAG was organized in 1960, soon after the first sit-ins in

Greensboro. Some of its initial demonstrations attracted as many as two hundred participants. One demonstration conducted during the summer of 1960 attracted five congressmen.

During its first year, NAG's members succeeded in desegregating about twenty-five facilities, including lunch counters, restaurants, a movie theater and Washington, D.C.'s only amusement park. At least one hundred persons were arrested in connection with demonstrations conducted by NAG.

The organization's name symbolized the determination of its members to "nag" the conscience of Washington. The name also reflected the theme of passive aggressive protest that characterized that stage of the civil rights movement.

By the time I joined NAG in the winter of 1962, most of Howard's students had lost interest in it. Picket lines and demonstrations were not considered glamorous activities anymore. The twenty-five to thirty students who belonged to the organization were considered "kooks."

NAG was a "Friends of SNCC" affiliate. This meant that those of us who belonged to it were unofficial members of SNCC. As members of NAG, we could attend some SNCC meetings and vote in some SNCC elections. There were scores of other Friends of SNCC groups on other campuses, especially in the North. Most of these organizations devoted their energies to fund-raising projects; few were actively involved in campus or community politics.

NAG was different. Although we sponsored dances and benefits for SNCC, that was not our primary task. Our primary task was demonstrating. We had a lot of good people, most of whom were Howard students—Courtland Cox, Murial Tillenez, Stokely, Stanley Wise, William (Bill) Mahoney, Ed Brown (Rap's brother) and Phil Hutchings.

Whenever black people in the city of Washington needed pickets, they would get in touch with NAG. It didn't matter to us if it was cold outside. If we thought we could help black people, we didn't mind demonstrating.

We frequently picketed various government departments. We weren't afraid of any of them. At different times the second semester, we picketed the Justice Department, Congress and the White House.

By the end of my second semester at Howard, I went on campus only to eat, sleep, attend classes and participate in periodic NAG rallies. I spent the rest of my time demonstrating and getting to know the people who lived in the huge black ghetto surrounding the campus. Unlike our classmates, the people in the community had a great deal of respect for those of us who belonged to NAG. We were great heroes to the young kids.

I spent a lot of time at Stokely's apartment. The crowded little cubicle was located a few blocks from campus. We considered it NAG's unofficial headquarters, and all the organization's members spent long hours there arguing politics and movement strategy. When it was time to eat and a good argument was in progress, we would pass a hat and send out for cold cuts.

NAG initiated an awareness program during the second semester. The program was designed to make people on the campus more conscious of the movement. We brought in several prominent speakers to augment our efforts, including Bayard Rustin. Although he had been rejected by SNCC's leaders two years before, we were enthusiastic when he accepted our invitation. The accumulated experience of two years had made a lot of difference in the political maturity of the student movement. We were all much more sophisticated and radical. If anyone had demanded that we deny our forum to Rustin in order to receive their support, we would have told them without hesitating to go to hell.

Rustin's address dealt with the state of the movement and the need for a coalition between blacks and organized labor. Although we did not share his trust in organized labor, we respected him. And after he finished speaking, we invited him over to Stokely's apartment for a rap session. To our proud surprise, he accepted our invitation.

At the end of my freshman year, I decided not to return to Denmark for the summer. I made the decision for two reasons: I did not want to live in my father's house any longer, and I felt I should remain in Washington, where a lot of important political things were happening. My decision was also affected by the fact that most of NAG's members had decided to do the same. Although my parents were hurt when I told them that I did not intend to spend the summer with them, they did not try to dissuade me.

I found a job washing pots in a Hot Shoppe restaurant. It was hard work and the pay was terrible. I didn't mind though. I was on my own, doing as I pleased.

Near the end of June, Bayard Rustin got in touch with NAG. He told us that plans were being made for a mammoth "March on Washington." The march, which was sponsored by SNCC, CORE, SCLC, the NAACP and the Urban League, was being conducted to emphasize the problems of poor blacks. It was to be a confrontation between black people and the federal bureaucracy. Rustin told us that some people were talking about disrupting Congress, picketing the White House, stopping service at bus and train stations and lying down on the runways at the airports.

Bayard, who was one of the march's deputy directors, said that thousands of people were expected. He wanted to know if he could count on NAG's members to assist the Planning Committee. We were honored.

"We're ready to do anything we can to help," we told him.

Bayard proved to be an extremely good organizer. He was on top of everything. He had us working all over town. Bill Mahoney was assigned to strategy. His job was developing tactics that would highlight certain aspects of the march. Other members of NAG worked on logistics. Ed Brown was assigned to transportation. I didn't have a specific job. Some days I was assigned to office jobs and other days I ran errands.

As the march date drew nearer, we began to suspect that something was amiss. Emphasis was being shifted from disruption and confrontation. We began to hear talk of the march as a coalition demonstration by workingmen—black and white. Moreover, several white liberals were brought in to act as cosponsors.

The march was being made "respectable." President Kennedy gave it his blessings. Then the Roman Catholic Church joined the bandwagon. On August 22, six days before the march, the Church's hierarchy issued a pastoral letter calling for all Catholics to become actively involved in assuring that "voting, jobs, housing, education and public facilities are freely available to every American."

We did not realize the extent to which the march had been co-opted until we saw the list of demands that the "official leaders" had drawn up. The vague demands, which were basically unrelated to the dominant thrust of the movement, were the same ones the Democratic party's liberal–labor coalition had been espousing for years:

> (1) Comprehensive and effective civil rights legislation from the present Congress, including provisions guaranteeing all Americans access to all public accommodations, adequate and integrated education, protection of the right to vote and decent housing.

(2) Withholding of federal funds from all programs in which discrimination existed.

(3) Desegregation of all school districts in 1963.

(4) Reduction of the congressional representation of states where citizens were disenfranchised.

(5) An Executive order banning discrimination in all housing supported by federal funds.

(6) Authority for the attorney general to institute injunctive suits when any constitutional right was violated.

(7) A massive federal program to train and place all unemployed workers, Negro and white, in meaningful jobs at decent wages.

(8) A national minimum wage of two dollars an hour.

(9) A broadened Fair Labor Standards Act to include all areas of employment now excluded.

(10) A federal Fair Employment Practices Act barring discrimination by federal, state and municipal governments, and by employers, contractors, employment agencies and trade unions.

Before we could protest the demands, we were faced with a bigger problem: the move to censor John Lewis's speech. John, who'd replaced Marion Barry as SNCC's chairman, had been working in the South on voter registration. And his speech reflected the militant concerns of the organization's members. When advance copies of it were distributed to leaders of the other sponsoring groups, they were very critical.

At an impromptu meeting called to discuss the speech, John was told that it was inconsistent with the spirit of the march. Several of those present at the meeting indicated that they particularly objected to his claim that President Kennedy's civil rights bill was "too little, too late." They also objected to his saying that SNCC's members intended to march through the South like Sherman and burn Jim Crow to the ground. Archbishop Cardinal O'Boyle let it be known that he would not give the invocation unless passages that he considered offensive were deleted from John's speech.

John, who'd been training for the ministry before he joined SNCC, was confused by his critics. He is mild-mannered and not prone to react in anger to any provocation. Although he felt that he was right, he was not absolutely certain that his critics were wrong. At the conclusion of the meeting, he gathered a group of SNCC and NAG members in order to get advice.

Our first impulse was to say, "The hell with them, John. Go ahead and give the speech the way you wrote it." After thinking about the situation, however, we realized that we were not in a position to do that.

SNCC was the smallest and least influential of the major civil rights organizations. And there was no way we could force the other organizations to let John give his original speech. More important, we did not want to make a big fuss and show the nation that there were serious divisions in our ranks. There was nothing we could do. John deleted those passages from his speech which offended the other leaders.

The public probably never would have heard about the censoring of John's speech if it had not been for Julian Bond's action. Because he was SNCC's publicity director, Julian was responsible for distributing copies of John's speech to the press. He gave every reporter a copy of both speeches and told them to look for the differences between the two. Julian admits that this was done without John's knowledge.

For me, the march was anticlimactic. I had seen too much. The censoring of John's speech, the changed emphasis, the decision not to disrupt—all these things were too fresh in my mind. I spent most of the long afternoon wandering aimlessly through the huge throng of 250,000 people: watching, listening and thinking.

I was very proud of John Lewis. Despite the deletions, his speech was very good:

> We march today for jobs and freedom, but we have nothing to be proud of, for hundreds and thousands of our brothers are not here—for they have no money for their transportation, for they are receiving starvation wages . . . or no wages at all.
>
> In good conscience, we cannot support the Administration's Civil Rights Bill. . . .
>
> This bill will not protect young children and old women from police dogs and fire hoses when engaging in peaceful demonstrations. This bill will not protect the citizens of Danville, Virginia, who must live in constant fear in a police state. This bill will not protect the hundreds of people who have been arrested on trumped-up charges like those in Americus, Georgia, where four young men are in jail, facing a death penalty, for engaging in peaceful protest. . . .

> I want to know—which side is the federal government on?
>
> The revolution is a serious one. Mr. Kennedy is trying to take the revolution out of the streets and put it in the courts. Listen, Mr. Kennedy, the black masses are on the march for jobs and freedom, and we must say to the politicians that there won't be a "cooling-off period."

When John asked, "Which side is the federal government on?" he exposed one of the most crucial issues in the movement. Federal officials had been procrastinating since the movement's inception in Greensboro. Several people had been beaten and others killed while federal officials took no action. It was time for the federal government to take a stand. President Kennedy's civil rights bill was beside the point. The demands of the march's "official leaders" were beside the point, too.

John was attempting to get the federal government to declare itself once and for all. It was essential that people working in the movement get an answer to John's question. Unfortunately, none of the other speakers followed through on the question.

Dr. King, who was introduced by A. Philip Randolph as "the moral leader of the nation," was the individual who profited most from the march. The reason for this is simple: his image was entirely consistent with the focus of the march. He was a dignified, moderate man who still believed in the goodness of the government and its operators.

Dr. King's speech was a thing of beauty. Despite my disillusionment with the whole affair, his rumbling voice and the eloquent simplicity of his "Dream" moved me. Like many of those standing near me in the crowd, I had tears in my eyes when he finished.

The people who got the most out of the march were the poor farmers and sharecroppers whom SNCC's organizers brought from Mississippi, Alabama and southwest Georgia. The march was a tremendous inspiration to them. It helped them believe that they were not alone, that there really were people in the nation who cared what happened to them.

CHAPTER 6

Small War in Cambridge

In late April, 1964, NAG received an urgent appeal for assistance from Gloria Richardson, leader of the black movement in Cambridge, Maryland.

"Governor Wallace is coming to town. He is scheduled to speak to an all-white audience in a public skating rink. We feel that his appearance must be challenged.

"The National Guard has been in Cambridge for almost a year—surrounding the black community every night. The people are getting discouraged. If we allow Wallace to come here to speak under these circumstances without a challenge, the local movement may die."

Stokely, myself, and several other NAG members went to Cambridge to help organize the anti-Wallace challenge. While there, we had an experience that nearly proved to be more than we could handle.

I knew when I decided to go that I was about to get involved in a struggle far more dangerous than anything I had previously experienced. I also knew that my decision would result in my flunking out of Howard. It was the middle of the semester and there was no way I could make up the work I would miss. However, my conscience was clear. I wasn't worried about grades, school or danger. I had already decided—sometime after the march on Washington—that the movement was my *first* priority.

Cambridge is a small town on Maryland's east shore. It had about fourteen thousand residents, one-third of them black. Although Maryland is often considered a Northern state, relations between blacks and whites there were quite similar to those that divide small Southern towns into hostile racial enclaves.

Local blacks were convinced that the town's whites were intent on destroying them. We were repeatedly warned to "stay off the streets after dark if you don't want to get shot!" The icy stares of hate convinced me that the warnings were justified.

The issues in the Cambridge movement transcended Governor Wallace's planned appearance. They encompassed a number of broad demands related to basic economic problems. Over 29 percent of the town's blacks were chronically unemployed, and more than one-third of those blacks who held jobs worked less than thirty weeks of each year. Two-thirds of all the families living in the town's huge black ghetto had incomes of less than three thousand dollars a year. And more than 60 percent of all the homes located in that area did not have hot water.

Black demands to improve these conditions had precipitated white violence. A number of people had been brutally beaten and snipers frequently fired into black homes. By the time we got to town, Cambridge's blacks had stopped extolling the virtues of passive resistance. Guns were carried as a matter of course and it was understood that they would be used in case of attack.

After only three days, we had completely adjusted to the local movement. If attacked, we intended to defend ourselves—"*by any means necessary!*"

Although I had never been in a situation where I had to carry a gun, I had no qualms about it. I didn't intend to shoot anyone unless I absolutely had to. But I decided when I accepted a gun that it was just as necessary to the work we were doing as stirring speeches, picket signs and marches against blatantly racist presidential candidates.

We had no trouble enlisting support for the demonstration. All the town's blacks knew who Governor Wallace was and what he stood for. They also knew who Gloria Richardson was and what she stood for. She was *the leader* of the Cambridge Nonviolent Action Committee (CNAC).

Everywhere we went we heard nothing but praise for her.

Gloria deserved every bit of the respect and admiration that she received. She was, in fact, one of the most outstanding leaders in the civil rights movement.

Gloria was not a large woman. She probably did not weigh more than one hundred pounds. Although she was attractive, she was not beautiful. Her soft eyes and feminine nature frequently misled those who did not know her well because they concealed her fierce pride and determination. By the time I had been in Cambridge a week, I was convinced that Gloria, who called herself a "radical revolutionary," was one of the most dynamic women I would ever meet.

We had a large mass meeting on May 11, the night Governor Wallace was scheduled to speak. The meeting was held at an Elks Lodge in the black community. There was an overflow crowd and many people had to stand outside in the street. I remained outside, wandering through the crowd during the meeting. Tension was high. Although they were listening to the speeches, I could tell that the people were thinking about the upcoming demonstration.

When the meeting adjourned at 7:30 P.M., the people inside filed out into the street and began to form a line. It was dark and I could barely see their faces. There wasn't much talking. Everyone seemed to be deep in thought. We knew that the national guardsmen were going to try to stop us from getting to the skating rink. We also knew that they were going to have to use force to accomplish that objective. After about thirty minutes, Gloria moved to the head of the line and we started out.

I joined the line of marchers about ten feet behind Gloria. I wanted to be close enough to the front to see everything. There were about six hundred people in the line, most of them black. Some tried to sing freedom songs, but most of us remained grimly silent.

I had spent most of the previous day surveying the neighborhood with the men who were serving as marshals for the march. The skating rink was located in a white area, the wrong place to get confused or lost on a night demonstration. We had double-checked all the entrances and exits of the skating rink. In case we got that far, we wanted to know how to get in and out in a hurry.

We marched about three blocks before we saw the national guardsmen. They were standing in the middle of the street that divided the black and white sections of town. A low murmur passed through our ranks.

I looked at Gloria. We still had time to turn around and return to the lodge. Her expression didn't change. She kept marching straight for the line of guardsmen. When we were within a half-block of the guardsmen, I noticed that they were armed with rifles. Each of the rifles had a bayonet attached to the end of its muzzle.

"Those mothers mean business," I murmured to Stokely, who was about five feet to my right.

When Gloria got to within ten feet of the guardsmen, their commander, General George W. Gelston, stepped forward. He knew Gloria and spoke directly to her.

"You'll have to stop here because you don't have a march permit," he barked. "I can't allow you to go any farther!"

Gloria, who had paused to listen to General Gelston, was obviously trying to decide what to do. While she stood thinking, I noticed that the guardsmen had a large light with them, the kind used to spot airplanes. I also noticed a crowd of three to four hundred whites milling around in the street behind them.

It was a crucial moment, the kind that can make or break a movement. We all understood that Gloria was the only one who could decide its outcome. If she had told us to return to the lodge, we would have done so, even though we would not have wanted to.

"I'm going through," she said.

Without waiting to find out how we would respond, she headed straight for the armed guardsmen. John Batiste, one of the SNCC members who had been assigned to Cambridge, ran forward to join her. He caught up with her just as she reached the guardsmen, and the two of them were immediately grabbed and placed under arrest. Before they could be led away, Khaleel Sayyed,* one of NAG's members, rushed forward and joined them. All three were hustled away before we had time to react.

I was angry. Without thinking, I stepped forward. One of the

---

* Khaleel Sayyed was one of the four black men convicted of plotting to bomb the Statue of Liberty in 1965.

guardsmen swung at my head with the butt of his rifle. I ducked and grabbed the gun just in time to keep from being beaten. The shit's on now, I thought. We stood there grunting and wrestling over the rifle while everyone else watched. After a few moments, I got into position to push and jump backward at the same time.

General Gelston, who had witnessed my struggle with the guardsman, rushed forward and yelled, "Arrest that man!" He was pointing at me. I considered trying to escape but realized that it was impossible. The crowd was packed too tightly behind me. There was only one thing for me to do. I went limp. Before the guardsmen could reach me, fifty marchers rushed forward and covered me with their bodies.

The guardsmen tried to pull the seething mass of people off me, but it was useless. Everyone in the pile grabbed someone else's arms or legs. There was a lot of scrambling. And each time one of the guardsmen bent over to try to remove someone, he was bitten or kicked. Furthermore, new people were joining the pile faster than the guardsmen could remove those already there.

When General Gelston realized that his men could not extricate me, he called them back. The guardsmen rejoined their ranks and began to don gas masks.

We were prepared for tear gas. We had received instructions on how to thwart it before leaving the lodge.

"Lie down on the ground and place a wet handkerchief over your nose," we were told. "The wet handkerchief will make it possible for you to breathe. The gas will rise over your head and float away."

While we were busily tying our dampened handkerchiefs to our faces, a guardsman dressed in a strange uniform headed toward us. He looked like an astronaut. His uniform was iridescent and gave off a soft, eerie glow. He was wearing a helmet. And there were a couple of big tanks strapped to his back. There was a tube leading from the tanks and he was holding the end of it, a long funnel, in his right hand.

"My damn goodness!" exclaimed one of the men lying atop me.

"That's the damnedest tear-gas gun I've ever seen!" replied a second man.

The second man was wrong. The strange contraption was not a gas gun. It was a converted flame thrower. The tanks on the guardsman's back did not contain tear gas, either. They were filled with some kind of nauseating gas. By the time we realized all this, it was too late.

The clear, acrid gas did not rise. It remained close to the ground like early-morning smog. We were covered with it. It was pouring out of the funnel in such large amounts that our clothes were dripping wet.

The gas made our wet handkerchiefs burn like fire. It also burned our nostrils. When we attempted to breathe out of our mouths in order to save our nostrils, the gas attacked the insides of our mouths and throats. My throat and stomach felt as if I had gulped a mouthful of burning acid.

The gas threw us into total confusion. We forgot about demonstrating, Governor Wallace and the skating rink. Everybody jumped up and started running. I took about fifteen steps and collapsed. When I came to about thirty seconds later, I was lying in the middle of the street, on my back like a dying cockroach. I was reminded of where I was when I heard someone yell, "Get up, Cleve! Here they come!"

I held my aching head up and tried to peer out of my burning, tearing eyes. The guardsmen were about thirty feet from me. They were moving forward shoulder to shoulder with their bayoneted rifles extended like spears. They grunted in unison before taking each step: "Ah-HUMP-CLUMP-Ah-HUMP-CLUMP-Ah-HUMP-CLUMP!"

It took all my energy to get up off the ground and stagger away. I could hear the guardsmen advancing behind me as I fled dizzily down the dark street: "Ah-HUMP-CLUMP-Ah-HUMP-CLUMP-Ah-HUMP-CLUMP!"

I ran about a half-block before the guardsmen began to fire their rifles. They were grunting, shuffling and firing in unison: "Ah-HUMP-CLUMP-CHOW! Ah-HUMP-CLUMP-CHOW! Ah-HUMP-CLUMP-CHOW!" My throat was on fire. My legs felt like rubber bands and my mind was hallucinating.

"Gotta make it back to our side of town, gotta make it back to our side of town, gotta make it back to our side of town. . . ." I kept repeating.

I staggered up to a fence. A group of black men were standing in front of it. They had leaned two boards against the fence and were frantically rolling people over it. Most of the people, many of whom were hysterical women, were too sick and confused to help themselves. Some were still vomiting and defecating.

I managed to clamber over the fence. I thought I had reached safety. I was wrong. I was running in the street again. The guardsmen were still behind me. They were still coming—and still shooting: "Ah-HUMP-CLUMP-CHOW! Ah-HUMP-CLUMP-CHOW! Ah-HUMP-CLUMP-CHOW!"

All of a sudden, the street was bathed in a bright light. In my confused state of mind, I paused to catch my breath and figure out what was happening. Looking back in the direction of the guardsmen, I discovered the source of the light. The guardsmen had turned on the searchlight and pointed it in our direction. It was blinding. All I could see were the silver bayonets of the advancing guardsmen.

I am certain that a lot of people would have been seriously injured if a small group of black men had not started shooting at the guardsmen in order to slow them down. It was like a scene from a Western movie. The men would run a few steps, crouch on one knee and fire; run a few steps, crouch on one knee and fire.

I ran to the CNAC office, which was filled with people. Most of them were too sick to talk. Stokely was the sickest of all. He was in terrible shape. Tears were flowing from his eyes, his stomach was still retching and he was only partially conscious. I tried to talk to him, to ask him how he felt, but he didn't even recognize me.

"Come on! We've gotta get him to a hospital before he chokes to death!" I yelled to a group of men standing nearby.

Grabbing Stokely by his underarms, we dragged him to a car. He was too weak to do anything except moan. We were outside the black community before I realized what we were doing. Oh my God, I thought. Here we are driving, black and unarmed, through a hostile white community during a race war. My head began to clear up—fast.

"Let's go to police headquarters," I told the driver. "Maybe we can get a squad car to escort us to the hospital."

There was a small group of patrolmen standing in front of police headquarters when we drove up. One of them stuck his head through the driver's window.

"What's the problem, boys?"

"We've got a very sick man here. We'd like to know if we can get a squad car to escort us to the hospital," I said.

I was holding Stokely's head in my lap. The patrolman leaned forward to get a closer look.

"Hell, ain't nothin' wrong with that niggah!" he replied.

After standing up and saying something to his fellow patrolmen, he stuck his head back into the window.

"Take him down to the fire station. They have an ambulance. They'll take him to the hospital."

We got our biggest surprise of the night when we pulled up in front of the fire station: it was full of state troopers. Before we could restart the car and drive away, two of them drew their guns and started walking toward us. They ordered us out of the car—"real slow."

We were marched over to a truck that was being guarded by a small group of national guardsmen. When they told us to get in, Cliff Vaughn, a SNCC photographer, resisted. "I haven't broken any law and I'm not going to get into that truck." That was a mistake. One of the guardsmen rammed a bayonet through his leg. Blood spurted everywhere.

"You niggahs must not know where you are!" said one of the troopers. "If you don't know where you are, I think that we can show you." He looked at each one of us. No one, including Cliff, said a word. When he ordered us into the truck the second time, we did as we were told.

I was the first one into the truck. Although it was dark and my eyes were still tearing, I sensed the presence of someone else. Moving toward the rear, I saw three figures, two men and a woman. It was Gloria, John Batiste and Khaleel.

"Well, I'll be damn!" said John.

"What is that foul smell?" asked Gloria.

I peeled off my gas-soaked shirt and handed it to her.

We had just finished telling them about the gassing and the shoot-out when the troopers returned for Stokely and Cliff. Stokely was still semiconscious.

"We got a couple of niggahs in here who need to go to the hospital," said one of the troopers to the ambulance attendant. He then proceeded to drag Stokely from the truck as if he were a bag of rotten potatoes.

Shortly after the ambulance left, a police squad car arrived. A young white college student was sitting in the rear seat. He was one of the few whites who took part in the march.

When the officers took him from the car, he went limp. That gave the angry troopers an excuse to work off a little steam. They grabbed him by his ankles and dragged him across the asphalt parking lot. By the time they got him to the steps of the fire station, his back looked like a piece of fresh hamburger.

We sat in the truck for a half hour before the troopers returned and transferred us to a bus, which then proceeded to a National Guard armory in Pikesville, a small town near Baltimore. When the bus pulled out, it was accompanied by four cars filled with state troopers. Moreover, there were four national guardsmen and two state troopers guarding us inside the bus. I wouldn't have tried to escape for a million dollars.

They kept us locked in the Pikesville armory for two days and then returned us to Cambridge, where we were locked in the city jail. The Cambridge Nonviolent Action Committee, which had been working on our release, bailed us out after two hours.

Shortly afterward, Stokely and I returned to Washington. The struggle wasn't over in Cambridge, but we had a new assignment: recruiting students for the Mississippi Summer Project. We were assigned to campuses in the Washington–Maryland area. Although we were not official SNCC members, we had taken on all the responsibilities of members.

By the end of May, 1964, we had recruited about twenty volunteers. Those anxious to get movement experience would come to Washington on Friday nights and spend the weekend working in the local SNCC office. We steered them into a number of projects: collecting bail bonds; local demonstrations; clothing, book and food drives and office work. On a couple of occasions, we took them to Cambridge to work with CNAC.

I gained a great deal of respect for Stokely during this period. He was a tireless worker and proved during our long conversations to be an astute political thinker. His most impressive ability was teaching. He had an uncanny sense of just what to say in order to explain complex relationships and problems to newcomers. By the same token, he could sit for hours discussing deep philosophical problems without becoming vague or redundant.

Stokely had majored in philosophy at Howard and was a precise thinker. This helped me a lot because I had a tendency to deal with problems on the basis of my feelings. He helped me develop into a better theoretical thinker. Neither of us felt at that time, however, that we had the solution to the problems facing poor blacks.

Little did I know at the time that in less than three years we would become designers of an analysis destined to precipitate a decisive change in the direction of the movement.

Before leaving for Oxford, Ohio, where SNCC was to conduct orientation sessions for the Mississippi volunteers, Stokely and I persuaded about fifteen of the student recruits to live with us. We wanted to help prepare them mentally for the work ahead.

The students, especially the girls, had a hard time adjusting to the crowded, one-bedroom apartment. They were forced to develop completely new attitudes toward privacy and personal possessions. We all slept on the floor in sleeping bags and had to stand in line in the mornings to go to the bathroom. We ate in shifts and everybody had to help with the cooking, cleaning and washing.

We spent the long, warm days sitting on the porch talking. Most of the discussions were led by Stokely or me. The object was to get the students to open up and deal with those hangups and idiosyncracies that might lead to conflict later on.

Many of the discussions dealt with things like conquering fear and death. Everyone was very concerned about coping with fear. Some of the students were certain that they would be paralyzed by fear once they got to Mississippi. Others were really bothered by their realization that there were people in Mississippi who would just as soon kill them as swat a fly.

"I don't know if I'm going to be able to take it," said one of the girls.

"The most dangerous thing I've ever done in my life was drive down the freeway at rush hour. How can I cope with the Ku Klux Klan? Where am I going to get the courage to go to Mississippi and actually try to register voters?"

We tried to help them answer these questions. I talked at great length about the South and how to survive in it. I also discussed the fears I'd had while working in Cambridge and how I had kept them under control. Stokely talked of experiences he'd had while working in Mississippi. We didn't manage to eliminate their fears, we hadn't eliminated all of our own. But we did help them to understand that they could do what had to be done.

Everyone had parent problems. Although some objected more than others, none of our parents were happy with our decision to go to Mississippi. Upon hearing the news, most parents hit the ceiling.

"Are you out of your mind? It's too dangerous. Let someone else do it," was the typical parental response.

Carol Martin and Doris Wilkerson were particularly concerned about their parents. We found out why one warm evening about a week before we set out for Oxford.

There were about twelve of us sitting on the stoop, including Carol and Doris, when two well-dressed black ladies walked up. "Are you Stokely Carmichael?" asked one of the ladies.

I was sitting near the front of the porch and she was talking to me.

"She wants to know if I'm Stokely Carmichael," I mimicked.

Stokely was sitting behind me. We were all laughing good-naturedly when the lady reached into her purse and pulled out a spike-heeled shoe. She was standing close enough to hit me on the top of the head. I stopped laughing.

Realizing that the ladies were angry, Stokely stepped forward and identified himself. The ladies told him that they had come for their daughters, Carol and Doris. The girls stepped forward and joined Stokely. I could tell that they were embarrassed and apprehensive.

"Go in there. Get your belongings and let's go," said one of the women.

Although upset, Carol and Doris stood their ground.

"We're not going with you, Mother," said Carol.

The girls and their mothers argued back and forth while the rest of us listened. When Stokely tried to act as a mediator, one of the women told him to "shut up" and mind his own business. After a few moments, the girls' mothers asked them to walk down the street where they could talk in privacy. They walked about a half-block before resuming the argument. We watched. The women did most of the talking.

When the mothers realized that their daughters were not going to abide by their wishes, they hailed a passing police car and told the officer that the girls were runaways. The officer informed the mothers that there was nothing he could do after he found out that the girls were eighteen, too old to be considered juveniles.

The girls returned to the porch while their mothers were talking to the policeman. Their mothers followed and issued the ultimate parental threat: "If you don't come home with us tonight, don't come home at all. If you go to Mississippi with *these* people, you can consider yourselves homeless!"

The girls stood fast. They were crying when their mothers left without them. We spent the remainder of the night trying to console them. Each of us knew exactly how they felt. Whether our parents understood or not, we had made our decision. We were going to Mississippi, the middle of the iceberg, to reclaim what Bob Moses had described as "the stone that the builders rejected."

CHAPTER 7

The Search

THE SOMBER BLACKNESS of the Mississippi night was fading into a deep purple when we got back to the house. Scrambling from the bed of the old pickup truck in which we'd been riding, we headed for the front door. It was imperative that we get inside before anyone saw us. I stumbled on the heels of the fellow in front of me as we walked single-file across the sagging front porch. By the time I closed the door behind me, the truck had disappeared in the darkness.

Inside, exhausted, we dragged ourselves to our sleeping bags. Although my shoes were coated with claylike mud, I was too tired to remove them. Despite my exhaustion, I couldn't sleep well. Each time one of the fellows sleeping on the floor beside me turned, my restless mind would jerk me awake. After an hour, I gave up on sleep.

Grabbing a chair, I moved over to the window and sat down. The sky was now pink. There was a little breeze, just enough to keep the deep green leaves on the magnolia tree in the front yard moving in tiny circles. A rooster was crowing somewhere in the distance. An old black man was leading a mule down the road in front of the house. On their way to work, they already looked tired.

While watching the sun edge over the horizon, I began to think back over the events of the past few days and nights. The drive from Washington to Oxford, Ohio, had been a lot of fun. SNCC provided us with two cars, and we had stopped at every other restaurant along the way to buy food.

Stokely, who had bragged before we left that he was a superior driver, provided us with the biggest laugh of the trip. He went to sleep at the wheel and drove down the middle of the grassy freeway median for almost a mile before we caught him and woke up the other people in his car.

"Hey, you guys had better get that nigger from behind the wheel before he takes you home to glory," I yelled to his passengers.

Stokely, who had veered back on the roadway just before we caught him, spent the remainder of the trip swearing that we had made up the story in order to "impugn" his reputation.

There were a lot of problems in Oxford, most of them stemming from the innocence of the white volunteers. When we tried to explain the rural South to them, they would nod as if they understood, but their eyes remained blank, uncomprehending. Many of them talked about Mississippi as if it were somehow the same as the romanticized scenes they had read about in *Gone with the Wind*.

Communication was further complicated by the fact that only about 135 of the 900 student volunteers were black. Feeling outnumbered and misunderstood, most of the black students withdrew. Bob Moses communicated best with the white students. Most of them had heard about the work he had been doing in Mississippi long before they got to Oxford. He was a *culture hero* to them and they talked about him all the time. By the time they had been in Oxford a couple of days, many of the white students had begun to emulate Bob's slow, thoughtful manner of speaking. Others rushed downtown and purchased bib overalls like his.

In all fairness, I must admit that Bob had almost the same effect on blacks. There was something about him, the manner in which he carried himself, that seemed to draw all of us to him. He had been where we were going. And more important, he had emerged as the kind of person we wanted to be. He was speaking to us that night when the news arrived.

"Our goals are limited. If we can go and come back alive, that is something. If you can go into Negro homes and just sit and talk, that will be a huge job. We're not thinking of integrating the lunch counters. The Negroes in Mississippi haven't the money to eat in those places anyway. They still don't dare go into the white half of the integrated bus terminals—they just weigh that against having their houses bombed or losing their jobs."

He was patiently explaining the scope of the Summer Project and the nature of the problems we would encounter. He knew Mississippi backward and forward. He picked his words carefully. During those moments when he stood before us deep in thought, we waited quietly.

"You may find some difficult limiting situations. If you were in a house which was under attack, and the owner was shot, and there were kids there, and you could take his gun to protect them—should you? I can't answer that. I don't think that anyone can answer that."

He paused again. Shifting slowly on the balls of his feet, he looked up and began to speak again.

"We've had discussions all winter about race hatred." The auditorium was completely quiet. We knew that he was about to get to the heart of a problem that could divide and destroy us during the dangerous months ahead.

"There is an analogy to *The Plague*, by Camus. The country isn't willing yet to admit it has the plague, but it pervades the whole society. We must discuss it openly and honestly, even with the danger that we get too analytical and tangled up. If we ignore it, it's going to blow up in our faces."

He had paused to gather his thoughts when one of SNCC's staff members sprang to the stage and whispered in his ear. A look of extreme concern spread slowly across his normally impassive face. I sensed that something terrible had happened. After a long moment, Bob told us what it was.

"Yesterday morning, three of our people left Meridian, Mississippi, to investigate a church-burning in Neshoba County. They haven't come back, and we haven't had any word from them. We spoke to John Doar in the Justice Department. He promised to order the FBI to act, but the local FBI still says that they have been given no authority."

An anguished murmur passed through the auditorium. Bob left the stage immediately and a slight woman whose face was furrowed with deep worry lines took his place. She told us the names of the three missing workers: James Chaney, Andrew Goodman and Michael Schwerner. Her name was Rita Schwerner. Michael was her husband.

Some wanted to believe that the three men were only missing. I had no such illusions. I was one of those who volunteered to search for their bodies.

Our search party was divided into four teams. Each team was given money for food and gas and a map with three alternate routes marked off. We knew that Mississippi's highway patrolmen were trying to keep civil rights workers out of the state, so we used different entry routes: they couldn't patrol all the roads.

We left Oxford at about three in the afternoon and arrived at the Mississippi border early the next morning. There was a big green sign on the side of the road adorned with cotton plants: "Welcome to the Magnolia State."

"The point of no return," I murmured to my team.

Each team had been instructed to phone the SNCC office in Meridian as soon as it entered the state. We had been instructed to call from SNCC's office in Holly Springs. There was a skeleton crew of three at the office when we arrived. They were expecting us.

"No word yet," they replied to our anxious eyes.

Three of the four teams arrived in Meridian before our one o'clock check-in time. Stokely and Charlie Cobb were late. We knew that they were in the state because they had called early in the morning from the SNCC office in Greenwood. Although some of us wanted to go out and look for them, we weren't allowed to.

We spent the afternoon being briefed on the three missing men.

Mickey Schwerner, who had arrived in Mississippi with his wife Rita on January 17, 1964, was the key figure. He was twenty-five years old and had done some work at Columbia University's School of Social Work. He was also deeply committed to the movement.

Mickey and Rita were members of CORE. Soon after they arrived in Mississippi they went to Meridian, where they set up a community center. The center became very popular with Meridian's young blacks, who loved to go there to read, relax and discuss politics.

Although Mickey and Rita had problems identical to those of every other civil rights worker attempting to crack a new town, they refused to be intimidated. They shrugged off the harassment of local police and tried to reason with those whites who criticized them.

In the spring of 1964, James Chaney, a twenty-one-year-old black high school dropout, joined the staff at the community center. He was from the Meridian area and wanted to do what he could to improve conditions for local blacks.

Mickey and James made a good team. Mickey had a good understanding of formal politics and shared his knowledge with James. Although he didn't know a great deal about formal politics, James understood the art of survival in Mississippi. He shared this knowledge with Mickey.

Mickey, Rita and James went to Oxford, Ohio, in mid-June to help run orientation sessions for the Summer Project. While there, they met Andrew Goodman. Like Mickey and Rita, Andrew was from New York. He was also a student at Queens College, Rita's alma mater. Although he was only twenty-one years old, he was mature and appeared to be a bit more confident than most of the student volunteers. Noticing these qualities, Rita, Mickey and James asked him to come to Meridian and work with them. Honored by their request, he accepted.

On Wednesday, June 17, Mickey got word that one of the churches he and James had been working in had been firebombed. Because they were concerned about the bombing's effect on the local people, Mickey and James decided to leave Oxford early and return to Meridian. Although Rita wanted to join them, they persuaded her to remain in Oxford for another week, training the volunteers.

On Saturday, June 20, Mickey, James and Andrew packed their car and headed south. It was three in the morning when they left. There was no way for Rita to know when she said good-bye that they would be dead in a few hours.

The three tired men arrived in Meridian at about 5 A.M. Sunday morning. After getting a couple of hours' sleep, they made plans to drive to Neshoba County and inspect the ruins of the church. Just before leaving, they instructed Louise Hermey, one of the girls working in the office, to call COFO headquarters in Jackson, the local police and the FBI if they were not back by 4 P.M.

That was the last time she saw them alive.

Louise, an attractive black girl, was one of the people in the Meridian office who briefed us. She was terribly upset and had a difficult time talking. She nearly burst into tears on several occasions. Although she still hoped that the missing men would be found alive, she *knew* what the rest of us knew: they were already dead.

Several times during the long briefing session, I walked to the window and looked down into the street in front of the office. Stokely and Charlie Cobb were still missing. When nightfall came and they still hadn't arrived, we began to wonder if they had met with the same fate as Mickey, James and Andrew.

Stokely and Charlie didn't show up until the next afternoon. They had been stopped in a notorious little town called Durant and charged with speeding. The police had arrested Stokely because he was driving. They had released Charlie but he insisted upon being locked up overnight; there was a crowd of hostile whites milling around in front of the jail. When Charlie was freed the following morning, he retrieved the money they had hidden beneath the floorboards of their car and paid Stokely's fine.

Shortly after Stokely and Charlie arrived, we drove over to Neshoba County and inspected the charred remains of the church that the three missing men had gone to investigate. It was a short distance from Meridian, so we all rode in one car. Although there were a few jokes about the girls' having to sit on the fellows' laps, we were mostly quiet.

The church had been one of those large old wooden structures that dot the countryside in the rural South. The cement blocks that had held it above the dusty ground were strewn about like the bones of some prehistoric creature. I was reminded while looking at them of the shotgun shanties in Denmark's ghetto across the tracks. They were propped up on the same kind of blocks.

The church had obviously been set afire, the smell of kerosene was still strong. Searching quietly through the oily ashes, we tried unsuccessfully to find some clue to the missing men. We gave up after forty-five minutes.

Getting back into the car, we drove a short distance to the house that was to serve as our hideout. It belonged to an old farmer whom I'll call Mr. Jones. His life would be in danger if I revealed his real name. The house had about five rooms, and, something unusual for black homes in that area, an indoor toilet. Water was drawn from a well with the help of an electric pump.

Mr. Jones, who lived with his wife, son and two daughters, was expecting us. His wife had prepared a big meal of greens, cornbread, buttermilk, candied yams and ham hocks. While we ate, they told us what they knew about the church-burning and the missing men.

"I believe that the peckerwoods burned the church and then killed them boys because us church folks was working with the COFO voter-registration people," Mr. Jones said.

"Ain't no telling where they done hid the bodies," his wife added.

"Y'all welcome to stay here and search as long as you want, though. We got everything set up for you."

Mr. Jones told us that he had organized a twenty-four-hour guard for the house. "I'll be sitting on the front porch with my shotgun every night and there'll be a man in the barn behind the house with a rifle."

We were told not to wander around the neighborhood at night because people were nervous and prone to shoot first and ask questions later. Just before leaving to check on some last-minute details, Mr. Jones told us that it would be safer for all concerned if we would remain in the house during the day. "The fewer people who knows that you all is here, the better," he said.

Soon after we finished eating, we drew a crude map of the area surrounding Philadelphia, the town where the missing men were last seen alive. The map was divided into sections and we made plans to make a systematic search of each one.

The routine we pursued during the next days and nights was harrowing. Every moment was filled with the possibility of discovery and death. Although we were supposed to sleep during the days, it was nearly impossible. We were too tense. The broiling heat and soupy humidity added to our discomfort.

One of the fellows had a small transistor radio and we would turn it on now and again to get news from the outside. We were listening to the radio one afternoon when I heard a local newsman refer to the missing men as "a group of race-mixin' outside agitators."

"It makes no difference to that son-of-a-bitch that James Chaney was born and raised in this goddam hell hole," I raged at the tiny radio.

Sometimes we saw copies of local white newspapers. All of them were against us. Most ran editorials suggesting that the missing men were hiding in order to give the state of Mississippi a bad image. The newspapers carried long, detailed stories on the preparations that law-enforcement officials had taken to thwart the Summer Project. The most extensive preparations had been made in Jackson.

Mayor Allen Thompson had increased his police force from 390 to 450, plus 2 horses and 6 dogs. Our police force is "twice as big as any city our size," the mayor boasted. The Jackson police department had purchased 200 new shotguns, stockpiled tear gas and issued gas masks to each patrolman. The department's motorized fleet included 3 canvas-canopied troop carriers, 2 half-ton searchlight trucks and 3 giant trailer trucks to transport demonstrators to 2 huge detention compounds.

"I think we can take care of 25,000," the mayor bragged to one reporter.

The pride of Mayor Thompson's militaristic police force was a monstrosity called "Thompson's Tank." The six-and-a-half-ton battle wagon contained shotguns, steel walls, bullet-proof windows, tear-gas guns and a submachine gun.

"They won't have a chance," crowed Mayor Thompson.

Fortunately, the paranoia of the mayor and his police officials was not one of our immediate problems.

The tension that kept us awake during the day took on a different character when the sun went down. The fear we felt during the day was diffuse. It produced restlessness and fatigue. The tension that accompanied the coming of night centered in the abdomen. It was precise and tended to heighten all the senses. At night I could hear, see, smell and feel hundreds of things that completely escaped my notice during the day.

Although SNCC and CORE were nonviolent organizations, we did not intend to remain nonviolent if apprehended. We knew that we would probably be tortured and killed if we were captured and our mission discovered. We did not intend to let that happen if we could help it.

Shortly before we left Mr. Jones's home on the first night, someone raised a question about guns. After discussing the pros and cons, we decided that guns were out. Too noisy. Feeling like a guerrilla in Vietnam or Latin America—at least I imagined that I felt as they do—I was prepared to kill if that was what it would take to keep me alive.

We were transported to the areas to be searched in an old pickup truck. Even though we never left the house until midnight or 1 A.M., it was too dangerous for us to travel the main roads. They were being patrolled by Klansmen and the local police. We traveled instead along the dusty, rut-filled back roads that bisect the cotton fields.

The driver of the truck, one of Mr. Jones's friends, never used his headlights. Whenever we approached a particularly hazardous section in the road, he would flick on his parking lights for a scant moment in order to get his bearings. He knew those unmarked back roads better than most people know the tendons on the backs of their hands.

I never ceased to be amazed at his skill. Speeding blindly along parallel ruts divided by a black strip of weeds and stray cotton plants, he would suddenly jerk the steering wheel to the right or left just in time to careen down another weed-choked trail. Had it not been for the aid of Mr. Jones, his family and his friends, we could not have survived for more than a few hours.

The assistance we received from a group of black share-croppers who lived in and around Philadelphia was crucial. Each day these men would leave home as if they were going hunting. Carrying shotguns and accompanied by long-eared hounds, they would spread out across the countryside searching for places where the three missing bodies might be buried. They reported likely places to the truck driver. He would relay their findings to us when he picked us up.

These men were native Mississippians who had no illusions about the perilous nature of their task. They *knew* that they were risking their lives and the lives of their wives and children by helping us.

"We gotta do it," one of the men told me one afternoon. "Somehow, it might make things better for the young kids. It's already too late for us old folks."

The procedure was always the same. Piling from the truck at 1:30 or 2 A.M., we would fan out. Walking slowly, and almost never talking—we searched swamps, creeks, old houses, abandoned barns, orchards, tangled underbrush and unused wells. Most of us used long sticks to probe the many ditches and holes we encountered. When the sticks proved inadequate, as they frequently did, we had to feel about in the dark with our hands and feet.

I spent a lot of time thinking about what I would do, how I would react, if we found the bodies. They had been dead long enough to begin decomposing. There was no telling what condition they would be in. Some wild animals, especially opossums, eat the flesh of dead animals.

"I just hope that whoever killed them had the decency to bury them," I said to the rest of the searchers one night while we were resting.

Our search was complicated by the poisonous snakes and spiders that abound in rural Mississippi. It was not unusual to step on a big, squirming snake or have a hairy-legged spider crawl down an open shirt collar. I still hate to think about the times I was poking around a tangled patch of prickly blackberry bushes only to be interrupted by the ominous hiss of an unseen snake. The hordes of mosquitoes and chiggers that plagued us were almost as troubling as the snakes and spiders.

Although flashlights would have been very helpful, we didn't dare use them. There was a fire tower nearby and the watchmen would have sent someone to investigate if they had seen any unusual lights. Our fears of being detected by those watchmen led us to discard all metal objects, including our belt buckles, which might reflect light. I kept my pants up with a section of Mrs. Jones's clothesline.

Sometimes we had to pass near farmhouses occupied by whites. This was extremely dangerous. Most farmers have watchdogs and we were always afraid that one of the dogs might detect us. Luckily, that never happened.

MY THOUGHTS WERE interrupted by a shuffling on the floor behind me. The other fellows were getting up. The sun was hovering low over the horizon and its dazzling rays indicated that we were in for another scorcher. Moving away from the window, I headed for the kitchen, where Mrs. Jones was cooking breakfast. Big clumps of dried mud flaked from my boots and preceded me across the floor.

We spent most of the morning trying to decide whether to continue the search. Even though we wanted to continue, we had begun to realize the situation was utterly hopeless. The unknown killers had made certain that no one was going to stumble over the bodies accidentally. Furthermore, our intelligence sources—the grapevine—indicated the Klan had gotten wind of our presence and had begun to search for us. Reluctantly, we decided to report back to regional headquarters in Meridian.

It was late in the afternoon when we arrived there. Sitting in the cluttered office, tired and depressed, we were informed of what had happened since we left Meridian. I listened, but the words really weren't registering. My mind was involved in a circular fit of depression. The summer hadn't even begun and we had already lost three good men.

CHAPTER 8

The Long, Hot Summer

IT WAS THE longest nightmare I have ever had: three months, June, July and August of 1964. I was nineteen years old, a man-child immersed in the seething core of the "long, hot summer." These grim statistics relate only a small portion of the horror: one thousand arrests, thirty-five shooting incidents, thirty homes and other buildings bombed; thirty-five churches burned, eighty beatings and at least six persons murdered. Although most of us managed to leave Mississippi, none of us escaped without terrible scars. It happened eight years ago, but the scars are still there, deep inside, where I suspect they will remain for the rest of our lives.

It began in earnest for me when I returned from Meridian to Holly Springs, Mississippi, which was to be our home base for the summer. A small group of people ran out of the office at 100 Rust Avenue when we drove up. They were summer volunteers, our coworkers. Most of them had just arrived from the Oxford orientation sessions. Their sad, searching eyes asked questions that would not be answered for many weeks: "Where are they? What happened to Chaney, Goodman and Schwerner?"

Ivanhoe was the Project director. Although he was of medium stature, standing about five ten and weighing no more than 175 pounds, he had tremendous presence and was a natural leader. He had been working with SNCC for more than two years and knew the South well. He'd made his first trip South in 1962 while a student at Michigan State in response to a SNCC appeal for food and clothing for a group of starving sharecroppers in Leflore, Mississippi. Ivanhoe and his companion Ben Taylor, a fellow student from Michigan State, had been arrested in Clarksdale and charged with the possession of dangerous narcotics. The "narcotics" were aspirin and vitamins.

The two of them remained in jail for eleven days under fifteen thousand dollars' bond until nationwide protests brought their release. Despite the threats of Clarksdale's police, who confiscated the truckload of supplies they had been transporting and warned them not to be caught in Mississippi again, Ivanhoe made twelve more trips from Michigan to Leflore, laden each time with badly needed food and supplies.

We had a long staff meeting on our first night in Holly Springs. Ivanhoe presided. Everyone sat quietly while he explained the long list of rules and regulations he had drawn up. Everyone was to be on his job by eight-thirty each morning; no one was to make any trip into the city or county without leaving his time of departure and expected return on the check-out list in the office; no one was to be out after dark unless he was on official business; all shades were to be pulled as soon as the sun went down, and no one was to make a target of himself by casting a shadow on a shade; local whites and the police were to be avoided whenever possible and never unnecessarily provoked.

Ivanhoe was particularly emphatic about affairs between blacks and whites. He told us that he did not intend to have any interracial relationships between staff members. In a very blunt and forceful manner, he told the white females that they were to avoid all romantic entanglements with local black males.

"Interracial relationships will provide local whites with the initiative they need to come in here and kill all of us. Even if the whites don't find out about them, the people will, and we won't be able to do anything afterwards to convince them that our primary interest here is political.

"Our entire effort will be negated if we lose the support and respect of the people. I don't intend for that to happen. Anyone who violates any of these rules will have to pack his bag and get his ass out of town. We're here to work. The time for bullshitting is past."

Pausing, Ivanhoe looked around the room from face to face. He wanted to make certain that everyone understood. We did.

The remainder of the evening was spent discussing COFO and the scope of the Summer Project. Although a number of organizations were supporting the Project, SNCC was the prime mover. SNCC people were coordinating four of the five Project areas—CORE was coordinating the fifth. SNCC was supplying 95 percent of the money for operating expenses and facilities throughout the state. Although Aaron Henry was president of COFO, Bob Moses was the Project director.

"The Summer Project has three major objectives," Ivanhoe explained, "registering voters, operating Freedom Schools and organizing Mississippi Freedom Democratic Party (MFDP) precincts. I intend for us to have the best goddamn project in the state. We're going to register more voters than anyone else, have the most efficient Freedom Schools and the best MFDP precincts!"

Before the meeting concluded, someone passed around a copy of a COFO publication titled "The General Condition of the Mississippi Negro." Despite my previous exposure to poverty and deprivation, I found it very disturbing.

The publication revealed that 66 percent of all Mississippi's blacks were living in "dilapidated or deteriorating" housing; that about one-third more blacks than whites died each year; that the chances of a black baby's dying in his first year were twice those of a white child; that one-half of the state's black schools had no equipment whatsoever; that more than 90 percent of the public libraries denied admission to blacks; that the state unemployment rate for blacks was twice that of whites and that the annual income rate for blacks was 71 percent less than that of whites.

"It really is going to be a long, hot summer," I muttered to no one in particular before tossing the pamphlet down and heading for bed.

I suspected from the way he conducted our first meeting that Ivanhoe was going to be a tough taskmaster. My suspicions were confirmed a couple of days later when Hardy Frye was arrested. Hardy, who had been assigned to register voters in the downtown area, was new and Ivanhoe told me to keep an eye on him.

I'd been watching Hardy for about a half hour when the chief of police walked up to him. "You're under arrest," the chief growled in a loud voice.

"Under arrest for what?" Hardy roared in an equally loud voice.

They were on the main street, which was moderately crowded with black and white shoppers. Their loud exchange caught the attention of everyone on the block. There must have been seventy people watching to see what would happen. This was the first confrontation between any of our staff members and local police officials. Everyone seemed to sense that it would have an important effect on the remainder of the summer.

"What the hell am I under arrest for? I ain't done a goddamn thing," Hardy insisted.

About five seven, Hardy was putting on a good show. Despite the fact that the police chief was several inches taller and probably fifty pounds heavier, Hardy wasn't backing down. He had been raised in Alabama and understood the symbolic importance of his confrontation with the chief.

The chief, who had obviously expected a less belligerent response, got confused. "You're under arrest for uh, uh, uhh, blocking traffic."

"Okay," said Hardy. "If I'm under arrest, let's go!"

With that, Hardy raised his hands high over his head and started walking toward the jail. The chief, who had reportedly killed several blacks, was further confused. Hardy was striding dramatically down the street as if on the way to the guillotine. He had a defiant smile playing at the corners of his mouth. The chief stumbled behind him, trying to catch up.

"Put your hands down, nigra!" he growled self-consciously.

When I got back to the office and told Ivanhoe what had happened, he jumped up from his desk and said, "Let's go!" Within an hour he had paid Hardy's fine and had him back in the office.

"You okay, Hardy?" asked Ivanhoe.

"Yeah, I'm okay. They didn't do anything but threaten me."

"Well, go back out there and get back to work," Ivanhoe commanded.

It was Hardy's turn to be confused. "Get back out there?"

"Yes, get back out there. There were a lot of people watching when you got arrested. We can't let them think that we are afraid.

You know that. Go right back to the spot where you were when you were arrested and continue to try to register people. Act as if nothing had happened."

"You're the boss," grunted Hardy, who was already on his way out the door. Although the police watched him closely, they did not rearrest him.

Ivanhoe's handling of the situation established two things: his position as the unquestioned leader of the Project and respect for our staff from the blacks of Holly Springs. If Ivanhoe had handled the situation any differently, if he had been softhearted and allowed Hardy to take the remainder of the day off, we probably would have had many more problems than we did during the rest of the summer.

Although I was the assistant Project director, I did not get to spend much time in the office. I was assigned to coordinate voter registration, Freedom Schools and MFDP organizing in Marshall County. The hardest part of my job was organizing Freedom Days. Every two or three weeks we would schedule a Freedom Day and try to get all the people we had contacted to come into town and try to register. Because numerous people had been killed on Freedom Days in other parts of the state, we had to work extra hard to keep the people from losing their nerve.

As each Freedom Day approached, I would move out of the crowded Holly Springs office and into the county so that I could be closer to the people. Working from sunrise long into the hot, humid nights, I had to keep reassuring the people that all hell was not going to break loose if they tried to register.

"You can do it! You can do it! I'll be with you every step of the way," I repeated again and again.

The Freedom Schools played a large part in the success of the Freedom Days. The children who came to the schools, many of whom were twelve and thirteen and still couldn't read or write, understood the meaning of freedom. All the lessons in the schools were tied to the need for blacks to stand up and demand the freedom that was rightfully theirs. The children would return home from the schools and badger their reluctant parents into going to town and registering for freedom.

By midsummer the periodic Freedom Days had evolved into something more than just "registration" days when blacks asked for the opportunity to get their names on the voter roles. The people saw them as opportunities to stand before their peers, white and black, and declare that freedom was something they intended to have. There is nothing so awe-inspiring as a middle-aged sharecropper trudging up the steps to the voter registrar's office clad in brogans, denim overalls and a freshly starched white shirt—his only one. I grew to love the Freedom Days. More than anything else, they provided the motivation that kept me going.

One of our most difficult tasks was getting people to attend the MFDP precinct meetings. Everyone knew that the three missing men were associated in some way with the party. They also knew that all those who attended the precinct meetings were letting themselves in for the same fate as the missing men. Although most people would not openly support the MFDP, we received invaluable assistance from local black ministers. They allowed us to address their congregations after Sunday morning worship service and urge their members to attend the meetings. By my third week on the job, I could deliver a pretty good sermon—in support of voter registration and the MFDP.

Although Marshall was supposed to be one of the most liberal counties in Mississippi, we had numerous run-ins with law-enforcement officials. I remember two in particular: an event that I call The Chase, and the death of Wayne Yancey.

The Chase took place late one Friday night. Three carloads of us were on our way back to Holly Springs after a big MFDP rally when we noticed that we were being followed by a police car. I was driving the lead car, a 1962 Volvo.

"Maybe we won't have to make a run for it. He doesn't seem to be interested in harassing us," I said to Ralph Featherstone, who was sitting in the front seat with me.

"I hope you're right," he muttered.

The police car followed us until we were in the middle of Oxford, Mississippi, before flagging down the two cars following me. I drove on for about four blocks before pulling to the side of the road to see if the other cars were going to be allowed to continue. When it became obvious that they were going to be detained, I swung the Volvo around in a wide U-turn and headed back.

By the time I got back and pulled in behind the two cars, the Marshall County sheriff and several other police cars had arrived. The sheriff, a short man who looked like a Bantam rooster, immediately turned the situation into a dangerous game.

While a fast-growing crowd of whites looked on, he called us from the cars one by one and attempted to humiliate us. The police officers looked on, chuckling and smiling.

"What yo' name, boy?"

"Cleveland Sellers."

"Where you from?"

"Denmark, South Carolina."

"What's a South Carolina nigra doing over heah in Mississippi? Ain't you South Carolina nigras got enough trouble without comin' over heah to Mississippi tryin' ta stir up our nigras?"

Although his remarks angered me, I remained impassive. I knew that he wanted me to give him an excuse to attack. It took everything in me to keep from spitting in his face and cursing him. His little eyes were shining and his lips were wet. He was really enjoying himself. So was the crowd around us. There were about 250 of them and they were cheering and howling like spectators at a bullfight.

"What's the matta, nigra? Cain't chew talk? Ever' time I turn on the television I sees one of you SNCC nigras talkin' 'bout how bad us white Mississippians is. What chew got ta say now?"

"I don't have anything to say," I replied in a dry voice, which was as devoid of emotion as I could make it.

"Git yo' slack ass back in tha cah! If'n I ketch you ovah heah ah-gin, um gonna puhsonally see to it that you leave in a pine box! Now git!"

Then turning to the other officers, the sheriff said, "He got a white girl in the back of that cah. Take him back and git huh ovah heah."

He was talking about Kathy Kunstler, whose father—William Kunstler—was one of the lawyers working with COFO. "Don't let him get to you," I whispered to her as she clambered from the car. Although the night air was cool, I noticed small beads of perspiration on her forehead.

The sheriff began by questioning her about her birthplace, occupation and residence. He was speaking in a loud voice so that everyone in the tightly packed crowd could hear. Very quickly, he descended to the level of all too many white minds in Mississippi.

"Which one of them coons is you fuckin'?" The crowd roared its approval of the question.

"Slut, I know you fuckin' them niggers. Why else would you be down heah? Which one is it? If you tell me the truth, I'll let you go. Which one is it?"

Although she was clearly frightened, Kathy did not break. Remaining calm, she spoke when it was appropriate and otherwise remained silent.

The crowd, which contained several tobacco-chewing drunks, was so engrossed in the exchange between Kathy and the sheriff that the rest of us were completely forgotten. That was good, because it provided us the opportunity to pass notes from car to car.

There was a federal courthouse across the street and we decided to make a run for its steps if the crowd attacked. There were several empty pop bottles in our car and we passed them to the occupants of the other cars. They were our only means of defense. Fortunately, we did not have to use them. The sheriff decided to let us go.

"Take your white whores and get the hell out of Oxford!" he yelled. "If'n I ketch any one of you heah again, um gonna see to it that you git a quick trip to hell!" He then told one of his deputies to escort us to the edge of town.

Just before the deputy headed back for the center of town, we noticed that we were being followed. Featherstone stuck his head out the window as we turned a corner. There were seventeen cars behind us.

"Oh shit," I muttered under my breath. Just at that moment the deputy sped past us, headed in the opposite direction. There was a big smile on his face.

"Our only chance is to run for it," said Featherstone.

He didn't have to repeat himself. Within seconds, we were speeding down the dark highway at 105 miles per hour. I was driving the last car and wasn't at all certain if the little Volvo could keep up with the big Plymouth Furies that Ivanhoe and Hardy were driving in front of me.

Featherstone stuck his head out the window again as we sped into a long, sloping curve. "There are twenty-one now," he said.

Before I had time to dwell on the significance of his report, we were faced with a new peril. Roadblock dead ahead. Ivanhoe, who was driving the lead car, responded immediately. The roadblock was set up in the right lane only. The left lane was open so that oncoming traffic could get through. Swerving into the left lane, Ivanhoe gunned past the startled little group of white men gathered to the right of the road. Before they could respond, Hardy and I did the same.

The only thing that saved us that night was luck. We drove the thirty miles from Oxford to Holly Springs as if we were Grand Prix racers. Our pursuers slowed down for bridges, sharp curves and small towns. We didn't. Hitting 105, we roared through the two small towns along the way with our horns blaring and our gas pedals on the floor.

Charlie Scales and Wayne Yancey were not as lucky. Wayne was killed one Sunday afternoon in mid-July in a two-car collision. We were sitting in the office trying to catch up on paper work when news of the accident arrived: "You folks better get down to the hospital. Two of your boys had a head-on wreck out on the highway and one of 'em is dead!"

When we arrived at the hospital, Wayne Yancey's body was lying in the rear of the big hearse that had been used as an ambulance. Although the accident had occurred more than an hour before, no doctor had examined him. His blood was seeping through the floor of the hearse and had formed a large dark puddle on the ground.

"Goddammit, what the hell is going on here?" Ivanhoe yelled to one of the white police officers standing nonchalantly in front of the hearse.

"You cain't move the body," drawled one of the officers. "Mayor's orders."

We rushed to the rear of the hearse, which was owned by Mr. Brittenum, the town's lone black mortician. One of Wayne's feet was hanging from one of the doors. It was obvious that his ankle had been broken. We moved closer and looked through the rear window. His face was badly mangled. And his body was bruised and torn by several deep cuts. Blood was everywhere.

I was immediately reminded of the magazine pictures of Emmett Till's corpse. Fighting to control my rage, I backed slowly away from the hearse.

"His head went through the windshield," I heard someone explain to a group of curious onlookers.

I didn't have time to mourn Wayne's death at that moment. We had another problem. Charlie Scales, who had been driving the car in which Wayne was riding, was still alive. Kathy Dahl, who had been with us at the office when the news arrived, was inside the hospital attending him. The white doctors and nurses at the hospital refused to help her. They said they were too busy.

Kathy, who was frantically trying to stem the bleeding of Charlie's wounds, came out briefly and told us that he was going to die unless he got additional assistance. Because we couldn't take a chance on sending him to another Mississippi hospital where he would certainly be refused assistance, we got Mr. Brittenum to agree to transport him to John Gaston Hospital in Memphis. We were preparing to enter the hospital and remove Charlie when we were informed that he was under arrest—for the murder of Wayne Yancey.

"They're trying to kill him, too," bellowed Hardy. In a blind rage, Hardy, Ivanhoe and I rushed the doors of the hospital. We didn't have a chance. There were almost twenty police officers in front of the doors. Ivanhoe sent for a lawyer, who managed after begging, threatening and pleading, to get the mayor to allow Mr.

Brittenum to transport Charlie to the hospital in Memphis. If it had not been for Kathy, who remained with Charlie in the rear of the hearse, he probably would have died then.

Immediately after the hearse left for Memphis, Hardy and Bob Fullilove and I were dispatched to the service station where Charlie and Wayne's car, in which the two men had been riding, was being held. We *had* to recover the MFDP membership forms in the car before they were confiscated by the police. We also wanted to inspect the car for signs of foul play.

The car, a brand-new 1964 Ford, was a total loss. It looked as if it had been hit by an explosion. Glass and blood were everywhere. The force of the collision had smashed the steering column back against the backrest of the front seat. We found one of Wayne's shoes on the floor in the rear.

We had just finished collecting the registration forms and examining the car when two white men came out of the service station and approached us. "Y'all friends of them two who was riding in this cah?" I shook my head affirmatively.

"Too bad you'all weren't in it wit' 'em!"

"Take that bullshit and ram it up your mother's ass," I yelled at the men, who were apparently mechanics. The three of us were advancing toward them, ready to lock nuts, when one of them placed a hand in a rear pocket. We stopped and backed slowly toward our car.

Jumping into the car, we quickly rolled up the windows.

"Let's get out of here," I yelled to Bob, who was sitting in the driver's seat. The white man, his hand still in his pocket, was still advancing toward us. "Let's go!" I yelled as the man began to remove his hand. I was certain that he had a gun. He had a badge instead.

"I'm a police officer. You're under arrest," he said to me.

"Fuck him! Let's go!" I yelled to Bob. But Bob froze. I was arrested for insulting a police officer and resisting arrest. SNCC paid my five-thousand-dollar bond just in time for me to get out of jail and attend Wayne's funeral.

Although we tried, we never did find out just how the collision had occurred. There was nothing at the scene we could use to determine what had happened. We questioned the black man who was driving the car Wayne and Charlie collided with, but he couldn't tell us much.

"I was knocked unconscious immediately," he said. "All I remember is that when my car came over the hill, there was this other car coming straight for me, in my lane."

Charlie couldn't tell us much either. He only remembered one thing. "I was thrown from the car and when I came to there was a white man standing over me. He bent over and said, 'Just lie still and be quiet or you'll get the same thing as your buddy!'"

Wayne's death had a tremendous effect on us. After getting crazy drunk and brooding for a week or so, we tried to pull ourselves together and "keep on keepin' on," but it was impossible. The weeks of tension and strain, coupled with Wayne's brutal death, could not be ignored.

Hate and viciousness seemed to be everywhere. We realized that the only thing keeping us from sharing Wayne's fate was dumb luck. Death could come at any time in any form: a bullet between the shoulder blades, a fire bomb in the night, a pistol whipping, a lynching. I had never experienced such tension and near-paralyzing fear.

The horror of it all was magnified when the FBI found the decomposed bodies of Mickey, James and Andrew. I don't remember what I was doing when the news arrived. It doesn't really matter. What I do remember is the excruciating pain it caused in my stomach. The pain, which remained for several days and nights, plagued my mind until it was impossible for me to rest or to forget what had been done to those three innocent men.

Wrapped in ice and plastic bags to protect them from the intense heat, the three bodies were taken to the University of Mississippi Medical Center in Jackson. Mickey and Andrew had been shot through the head once each with a .38 caliber bullet. James Chaney had been shot three times and, according to one of the examining pathologists, brutally beaten.

"In my twenty-five years as a pathologist, I have never witnessed bones so severely shattered," said Dr. David Spain of New York after examining Chaney's body at his mother's request.

The bodies were found beneath an earthen dam on a farm three miles from Philadelphia. The farm belonged to a white trucker, Olen Burrage, who claimed that he did not know how the bodies had gotten beneath the dam. Herman Tucker, who had been paid $1,430 by Mr. Burrage to build the dam over the site early in the summer, pleaded ignorance.

"I don't know nothing about it," he said, "don't care nothing about it and don't want to discuss it."

Rita Schwerner, who was working for CORE in Washington when the bodies were found, uttered a terse comment when newsmen inquired about her feelings: "Three good men were killed—three good men who could have done a great deal for their country."

When asked if she thought something positive might come from the triple assassination, Mrs. Schwerner gave the only answer any of us could: "That is up to the people of the United States."

We approached our work with additional dedication and purpose. Despite our pain, we were determined to accomplish our goals. Working doggedly from sunup to sundown, we set about the task of organizing precinct, district and state MFDP meetings. We had a good case and we intended to take it to the Democratic National Convention in Atlantic City. We would present it to the party in power, to the elected representatives of "the people of the United States."

Not having the money to fly, we set out for Atlantic City in bus and automobile caravans. We knew that there was going to be resistance to our demands, but we were prepared. We took numerous items to be used as symbols and proof of the systematic oppression of Mississippi's blacks. They included notarized depositions from victims of discrimination, pictures of the conditions in which blacks were forced to live, reports on the economic deprivation of Mississippi blacks and a list of all the churches that had been burned or bombed. We also took the car in which Chaney, Goodman and Schwerner were riding on the night they were assassinated.

We would prove that the MFDP's delegates were more loyal to the stated goals of the national party than the state's regular delegation and, therefore, deserved to be seated in its place. We were scrupulous in our attention to detail. There was no way for anyone to claim that the MFDP's delegates were not legally and morally entitled to the seats. It was time for the national party, as they say in the black ghettos, "to shit or get off the pot."

We were thinking far beyond Atlantic City. If our venture there was successful, we intended to utilize similar tactics in other Southern states, particularly Georgia and South Carolina. Our ultimate goal was the destruction of the awesome power of the Dixiecrats, who controlled over 75 percent of the most important committees in Congress. With the Dixiecrats deposed, the way would have been clear for a wide-ranging redistribution of wealth, power and priorities throughout the nation.

This strategy, which to a large extent was the brainchild of Bob Moses, could have proven successful if, and only if, the leaders of the national party had been willing to support the MFDP challenge. As it was, they ignored principle and offered the MFDP delegates a compromise—two nonvoting seats beside the regular delegation.

The MFDP's delegates held a dramatic meeting at a local church in order to decide whether or not to accept the compromise. Although the press was excluded, several liberals, black and white, were in attendance. Most of the liberals, including Dr. King, Roy Wilkins, Whitney Young and Bayard Rustin, were in favor of accepting the compromise. They convinced some of the MFDP delegates to go along with their reasoning.

Bob Moses gave a short address. He told the delegates that SNCC's members had decided to remain out of the controversy and allow the delegates to make their own decision. Although we were hoping that they would reject the compromise, we didn't feel we had the right to ask them to do so. We realized that the two nonvoting seats were a reward of sorts. If they accepted them, the delegates could return to Mississippi with the realization that they had accomplished *something*. We weren't in a position to offer them anything.

Mrs. Fannie Lou Hamer, an eloquent black woman from Sunflower County, was the last person to address the meeting. Speaking slowly, but with great clarity, she brought it all home. She told the delegates that she had come to Atlantic City to unseat the regular delegation. Addressing the liberals, she said that she understood why they believed that two seats were better than none. Then she told them why they were wrong. She was on the verge of tears when she finished. Some of the rest of us were too.

Events after Mrs. Hamer's address are a matter of public record.

The MFDP's delegates informed the Democratic National Committee that they would not accept the compromise; the regular delegates from Mississippi were seated; Hubert H. Humphrey, who served as Lyndon Johnson's front man in support of the compromise, became the vice-presidential candidate.

We took note of these happenings, packed our exhibits and returned to Mississippi—proud and unbowed.

CHAPTER 9

A River of No Return

THE NATIONAL DEMOCRATIC party's rejection of the MFDP at the 1964 convention was to the civil rights movement what the Civil War was to American history: afterward, things could never be the same. Never again were we lulled into believing that our task was exposing injustices so that the "good" people of America could eliminate them. We left Atlantic City with the knowledge that the movement had turned into something else. After Atlantic City, our struggle was not for civil rights, but for liberation.

I did not return directly to Mississippi after the convention. I went instead to my parents' home in Denmark in order to rest and recuperate. The summer had been physically and psychologically punishing. I knew that if I returned to Mississippi before getting myself together, I would be dangerous to myself and to my coworkers. I just didn't have the strength to deal responsibly with another crisis like the death of Wayne Yancey.

While driving back to Mississippi across the farmlands of Georgia and Alabama, I reflected on the tense, halting conversations I'd had with my family during the short visit. Although I was only nineteen years old, I had seen so much more than they, been through so much. There was little common ground for communication.

I had thought I could discuss my experiences with Gwendolyn, but I couldn't. She was not a part of the movement and had no real understanding of what I'd been through. She was sympathetic, but I wasn't looking for sympathy. Though neither of us brought them up, all the problems that existed between me and my father before I went away to Howard were still there. I realized after the first couple of days at home that the only things we could safely discuss were the athletic teams at Voorhees and SCAT.

I found the discussions with my mother to be easiest. There was, however, a gulf between us, too. I wanted to bridge it, but couldn't. She wanted to know so many things: why I *had* to return to Mississippi, how much longer I intended to remain out of school, how long I intended to remain active in the movement?

"We worry about you so much. Why don't you write more often? Are you taking care of yourself? You seem to have lost weight."

What could I say? Certainly I had lost weight. Yes, I was taking care of myself—as best I could. Yes, I did intend to return to school—someday. Mississippi? I have to return because, because . . . It's something I must do. For the first time in my life, I understood how soldiers feel when they return from wars and have to grope unwillingly for answers to the terribly innocent "How was it?" questions of family and friends.

As my car sped across the state line of Mississippi, I thought about my last words with my father. I was sitting in the car in the driveway getting ready to leave when he stepped forward and extended his right hand. I grasped it and looked into his eyes.

Just for a moment, there was a look of tenderness and pain. I remembered that look from when I was a small boy. Before I could say anything, he caught himself and the look was gone, replaced by an invisible shield. His callus-covered hand, which I was still clutching for what may have been the last time, was warm and firm.

"Take care of yourself, boy," he whispered in a hoarse, breaking voice.

I didn't reply because I was fighting to hold back the tears.

It was dark when I pulled up in front of the Holly Springs office. Somehow, it looked smaller than before. The worn wooden steps and the door, which was emblazoned with a huge "Freedom Now" poster, looked years older. Climbing slowly from the car, I headed inside. Despite the considerable amount of sleep and rest I had gotten at my parents' home, I was still very tired.

"Hello, *Mr. Project Director,*" someone said when I entered the door.

The greeting reminded me of my new responsibilities. Ivanhoe had been transferred to SNCC's National Office in Atlanta. His new title was Administrative Assistant. He had been assigned the task of making the organization's wheels function more efficiently. Because I had been his assistant and knew the Holly Springs project from top to bottom, I was promoted to his old post.

The winter turned out to be just what I had expected: long, cold and hard. The picnic was over. The summer volunteers, the newspaper and television reporters and the local hangers-on were gone. For those of us who remained, there was only fear, work and exhaustion. The chilling winds, which swept across the barren fields and down the dusty streets, made our work all the more difficult.

One of our most important staff meetings was held in Waveland, Mississippi, in late November. A couple of weeks before the three-day meeting, the National Office in Atlanta sent out a leaflet requesting that interested parties submit position papers on topics that concerned them. The papers were to be used as discussion guidelines. The leaflet also contained a list of some of the most important questions to be discussed during the meeting: "Why do certain people have power? Whose interests are represented and how are they represented by the election of congressmen, the President? By the selection of Cabinet members and federal judges? How do the federal government, big business and labor unions relate to the Black Belt power structure on the issue of civil rights? What influences the relationship? How can an organization like SNCC help people participate in decision-making? How does an organization like SNCC structure itself to do the job?"

The position papers were passed out shortly before the meeting began. I took the time to read each one. Some were written extremely well. They were logical and the style indicated that their authors were former students. Others were written in painful prose, indicating their authors' lack of formal schooling.

The longest papers covered as many as ten pages. The shortest ones took up as little as one page. Regardless of style or length, all of them reflected the sincere desire of SNCC's members to deal honestly and effectively with the problems before us.

I remember one of the papers in particular. Like most of the others, it was unsigned. It was short and contained several grammatical and typographical errors. Nevertheless, it was one of the best papers submitted. I still remember the section in which the author attempted to sum up our dilemma: "We are," he wrote, "on a boat in the middle of the ocean. It has to be rebuilt in order to stay afloat. It also has to stay afloat in order to be rebuilt. Our problem is like that. Since we are out on the ocean we have to do it ourselves."

Jim Forman, SNCC's executive secretary, gave an outstanding speech at the beginning of the meeting. The speech was his attempt to provide a constructive framework for the inevitable arguments and disagreements that were to occur during the next three days. While listening to him move patiently from issue to issue, I began to realize that, in his own way, he was an orator. Unlike Dr. King, he does not have a great voice. His ability to communicate is derived from his great determination. The meeting hall was very quiet as he approached the end of his speech.

"Someone has written," he said, "that we are a boat afloat, and that it must be changed in order to stay afloat, and that it must stay afloat in order to be changed. There are some who say they don't understand that metaphor. Well, let me further confuse the picture by saying that we are on a river of no return."

It was a very productive conference. We returned to our various outposts afterward with a better sense of who we were and what we were trying to accomplish. We didn't answer all the questions, it was impossible to do that. But we dealt constructively with most of them. Our most important conclusion was that it was not possible for poor black people to solve their problems within the structure of the two-party system. We agreed that there was a need for alternative or parallel political structures.

Two major campaigns grew out of the conference: the Mississippi Challenge and the Selma to Montgomery March.

We launched the Challenge in January, 1965. It was based on Article I, Section 5, of the United States Constitution, which provides that the House of Representatives "shall be the judge of the elections, returns and qualifications of its own members." In accordance with procedures determined in Title 2 of the U. S.

Code, we prepared massive documentation to prove that the five Mississippi congressmen elected in November should not be seated.

We had irrefutable proof of the systematic manner in which Mississippi's blacks were excluded from participation in the electoral process. The Second Congressional District, which included Holly Springs, was typical. Of the adult population in the district, 52.4 percent was black. Yet only 2.9 percent of the blacks had been permitted to register and vote.

We contended that Representative Jamie L. Whitten, who had been elected to Congress in the November election from the Holly Springs District, and the four other Mississippi congressmen who were elected under similar circumstances, should not be seated. As was the case in Atlantic City, we had the law, massive evidence and morality on our side. When Congress opened on January 4, it had only to vote in accordance with its own rules. Of course, it did not do that.

When a roll call vote was taken on whether to deny seats to the five congressmen or to seat them until the charges against them could be investigated, 228 congressmen voted to seat them and 143 voted to keep them seatless. Although we were mildly surprised by the number of congressmen who supported the Challenge, the outcome of the vote was expected.

Our objectives in pursuing the Challenge were quite different from those we had had in Atlantic City. We launched the Challenge in order to prove that the system would not work for poor black people. Our aim was to disprove the notion that black people, if they play by the rules, can acquire equal justice under the law. In that sense, the Challenge was successful.

On January 2, 1965, two days before Congress voted on the Challenge, Dr. King announced to the press that he intended to lead a large voter-registration campaign in Selma, Alabama. I was not happy when I heard the news. I *knew* that the campaign was going to lead to a major confrontation between SNCC and SCLC. This was true for two reasons: SNCC already had a voter-registration staff in Selma; and SCLC and SNCC had completely different approaches to community organizing.

Four SNCC organizers, Bernard and Colia Lafayette, Frank Holloway and James Austin, had gone to Selma in the fall of 1962. The project they started had been in continuous operation. As was the case with SNCC projects in other areas, emphasis was placed on the development of grass roots organizations headed by local people. We called it participatory democracy: local people working to develop the power to control the significant events that affected their lives. Working, eating, sleeping, worshiping and organizing among the people—for years, if necessary—that was the SNCC way.

SCLC's approach was radically different. Instead of remaining in a community to develop autonomous organizations, it would organize dramatic demonstrations calculated to get the attention of the nation. After getting that attention, as was the case in Saint Augustine; Albany, Georgia; and Birmingham, SCLC would submit a list of demands to the local power structure, win minor concessions, proclaim a great moral victory and leave town.

The differences between the two approaches reflected the different analyses made by members of the two organizations. SNCC's members were convinced that black people would be free only when they took their destiny in their *own* hands and forced a change in the status quo.

SCLC's members, on the other hand, believed that blacks would be free when the federal government took steps to ensure that their rights were not violated. SCLC organized huge marches and demonstrations to activate public opinion in support of its goals. The thinking was that mobilized public opinion would result in civil rights legislation and federal action.

This approach invariably had a disastrous effect on the local movement in the communities in which SCLC worked. Although they would be lifted and inspired by the drama surrounding the giant demonstrations, local blacks would have little to show for their sacrifices when SCLC's entourage left town.

On January 18, less than three weeks after Dr. King announced his intention to go to Selma, SCLC launched a series of daily demonstrations that focused on police brutality and the local voter registrar, who required prospective black voters to pass an extremely difficult "literacy test."

Selma's police, who were led by a Bull Connor–like chief named Jim Clark, frequently used electric cattle prods, long nightsticks, chains and horses to harass the demonstrators. The attitude of the police and Selma's white power structure was symbolized by the button that Sheriff Clark took to wearing on his lapel. It contained one word, "Never!"

It became apparent soon after the demonstrations began that SNCC's organizers were not going to be able to continue their work. The huge SCLC demonstrations were too disruptive. The entire black community was mesmerized. Those who were not actively supporting the demonstrations spent most of their time discussing them. More important, few people had the energy to sit through long planning sessions with SNCC's organizers after having spent the entire day demonstrating with SCLC.

SNCC called a meeting to discuss the problem. The majority of those who attended agreed that the demonstrations were counterproductive. "They make the people feel very good, but they don't really accomplish very much," reported Stokely, who had gone to

Selma to observe. Because we did not want to denounce the demonstrations openly and give credence to the rumor that there was a major split between SCLC and SNCC, we decided that those organizers stationed in Selma should take a back seat.

"We, SNCC, will not actively support the demonstrations," Jim Forman said. "We will not try, however, to keep those local people who want to join them from doing so. When SCLC leaves town, we will attempt to pick up the pieces and work from there. Does everyone agree?"

At least one person didn't—John Lewis, SNCC'c chairman. John supported the demonstrations and was convinced that SNCC should do likewise. He was a member of SCLC's board of directors and had sat in during the planning sessions for the Selma campaign. He believed that the campaign was sound. He told us that he was a native Alabaman and that the demonstrations were good for Alabama's blacks. He also told us that he intended to support the demonstrations with or without SNCC support.

"If I can't participate as the chairman of SNCC," John declared, "I'll participate as a private individual."

A big argument ensued. Most of us felt that John was wrong. If he supported the demonstrations, many people would infer that SNCC also supported them. The fact that he was the *only* SNCC member doing so would be lost. John was convinced that he was right, however, and there was nothing we could say to change his mind.

The meeting ended with us agreeing that John had the right to go to Selma and to participate in the demonstrations—*as an individual.* If SNCC had been a different kind of organization, he probably would not have been allowed to do so. As it was, we supported the principle that every SNCC member had the right to comply with his conscience. There were many times during the next weeks when we sincerely regretted that decision.

On March 7, 1965, I was attending a statewide MFDP meeting in Jackson when I received an emergency call from the SNCC office in Selma.

"Get as many people as you can and get over here immediately. John Lewis and Hosea Williams of SCLC were leading a march and the police attacked them. A lot of people were hurt. John is in the hospital and we think he has a fractured skull."

In less than an hour, we assembled a crew, gassed up, packed extra clothes and sleeping bags. We had five cars, five people in each.

When we arrived in Selma, there was total chaos. At Brown's Chapel, the church where most of the demonstrations began, we witnessed a scene quite similar to the one in the Cambridge Nonviolent Action Committee office after the National Guard attack. The police officers had totally disoriented the people. The little church was filled with hysterical women, weeping children and thoroughly frustrated men. Many of the people still smelled of tear gas. Several wore makeshift bandages soaked through in spots with blood. SCLC officials were trying unsuccessfully to keep order.

Things were calmer at SNCC headquarters. There were only a few people there, most of them movement veterans. They had seen similar attacks in the past. They were still angry, however. I could tell by the intensity in their voices, the feel of the warm air, that the spirit was building. A major confrontation was in progress.

Because I realized that we were going to be in Selma for some time, I spent the remainder of the night reading through staff reports and SNCC research papers. I wanted to be thoroughly familiar with Selma and the course of the SCLC campaign.

As I read through the reports, I found that Selma was the

center of a struggle taking place between blacks and whites in nineteen Alabama counties described as the Black Belt because of their heavy black population. In Dallas County, of which Selma was the capital, 57 percent of the residents were black. Yet, only .9 percent of them were registered. In the adjoining counties, Lowndes and Wilcox, where the total black population exceeded fourteen thousand, not one black was registered.

The SCLC campaign to register black voters had experienced rough going from the beginning. On January 18, John Lewis and Dr. King led a demonstration of five hundred blacks to the courthouse. They were forced to stand in a debris-filled alley all day, and then no one was registered. While they were huddled in the alley, a white man who attempted to attack Dr. King had to be subdued by John Lewis. The essentially unproductive character of that day's events was typical of the next two months.

Wary of the increasingly violent encounters in Selma, SNCC's representatives had called for a meeting with SCLC's leaders on March 5. At the meeting, the people from SNCC expressed their opposition to SCLC's proposed march from Selma to Montgomery. They argued that the danger involved was greater than any possible achievements. SCLC's officials, supported by an adamant John Lewis, disagreed. They refused to cancel the march.

SNCC conducted an Executive Committee meeting the next day to discuss Selma and the proposed march. The consensus was that SNCC had an obligation to do whatever it could to support those local people who joined the march. Thus, it was decided that SNCC would provide to the marchers those services already agreed upon—radios, telephones and medical assistance.

And on March 7, John Lewis and Hosea Williams led the march that Governor Wallace had already instructed state troopers to stop by use of "every necessary measure."

The marchers started from Brown's Chapel late in the afternoon, five hundred people, walking two abreast. Singing lustily, the marchers approached the Edmund Pettus Bridge, which led out of town toward Montgomery. The singing grew louder as the people noticed Sheriff Clark's deputies stationed on each side of the narrow bridge.

After pausing a short moment, the head of the line moved across the bridge toward the line of Alabama state troopers stationed on the other side. When the marchers were less than fifty feet from the troopers, Major John Cloud shouted that they had two minutes to disperse. When Hosea sought to have a word with Major Cloud, he was refused.

"There is no word to be had."

While John and Hosea were trying to decide what to do, the state troopers, joined by Sheriff Clark's deputies, surged forward and attacked—with clubs, chains and electric cattle prods. Totally unprepared for the brutal attack, the people turned and amidst curses, yells and cries of pain, tried to escape. Lafayette Surney, a SNCC member who observed the scene from a nearby phone booth, delivered a minute-by-minute account over the phone to the SNCC office.

> 4:15 P.M. State Troopers are throwing tear gas at the people. A few are running back. A few are being blinded by tear gas. Somebody has been hurt—I don't know who. They're beating them and throwing tear gas at them.

> 4:16 P.M. Police are beating people on the streets. Oh, man, they're just picking them up and putting them in ambulances. People are getting hurt, pretty bad. There

> were two people on the ground in pretty bad shape. . . . I'm going to leave in a few minutes. . . .
>
> 4:17 P.M. Ambulances are going by with the sirens going. People are running, crying, telling what's happening.
>
> 4:18 P.M. Police are rushing people into alleys. I don't know why. People are screaming, hollering. They're bringing in more ambulances. People are running, hollering, crying. . . .
>
> 4:20 P.M. Here come the white hoodlums. I'm on the corner of one of the main streets. One lady screamed, "They're trying to kill me!"
>
> 4:26 P.M. They're going back to the church. I'm going too. . . .

The attack was responsible for a SNCC decision that would not have been made under normal circumstances. Early Monday morning, we decided to ignore our previous reservations and support a march to Montgomery. We were angry. And we wanted to show Governor Wallace, the Alabama State Highway Patrol, Sheriff Clark, Selma's whites, the federal government and poor Southern blacks in other Selmas that we didn't intend to take any more shit. We would ram the march down the throat of anyone who tried to stop us. We were ready to go—that afternoon.

As we should have expected, but didn't, Dr. King and his associates were of a different frame of mind. They argued that the march should be postponed until they could acquire a court order authorizing it. We told them that a court order just wasn't necessary. The local people were ready to march on Montgomery and we saw no reason to ask them to wait. While SCLC's officials were searching vainly for a judge to authorize the march, Willie Ricks, one of SNCC's most volatile members, instigated a series of impromptu marches in front of Brown's Chapel.

There were state troopers stationed at each end of the block. Their orders were to stop *all* marches. Ricks, a fiery speaker, got together a large crowd of schoolchildren and marched them back and forth between the two lines of nervous troopers. Each time the students turned and moved down the block they became more boisterous.

Ricks's purpose was obvious. He was trying to get the crowd to break through the line of troopers. The students proceeded to get into several shoving matches with the troopers. When Ricks realized that the students were not going to be able to overrun the troopers, he stood on the steps of the chapel and began suggesting that they disperse and reassemble downtown in front of the courthouse.

Ricks's actions infuriated Dr. King's associates. They were concerned that the impromptu marches would lead to additional violence. They attempted to talk to Ricks, to stop him from "inciting the people." But he wouldn't listen. When they tried to get us to stop him, we explained that we permitted SNCC's individual members to do whatever their consciences told them to do.

Certainly, we could have persuaded Ricks to withdraw if we had wanted to, but we didn't. We were convinced that the ban on marches was levied in order to break the momentum of the campaign. A lot of people in the community felt the same way.

"I want to march," said a middle-aged man in the crowd gathered before the church. "If they gonna be violence, that's okay. They caught us by surprise yesterday. Next time, we gonna give as good as we git."

SCLC had been faced with similar crises before. As a matter of fact, Dr. King had been forced on numerous occasions, including the night his own home was bombed during the Montgomery bus boycott in 1956, to plead for nonviolence in order to restrain blacks ready to defend themselves. Because they realized that Ricks's actions were exposing their reluctance to move *with* the people instead of *in front* of them, SCLC's officials called for a closed meeting with officials from SNCC.

The meeting was held in a back room of Brown's Chapel. Although those of us who belonged to SNCC were generally younger than Dr. King and his associates, we did not feel that we had to defer to them—on any level. We were "veterans" who had always held up our end of the struggle. We had paid our dues and had the scars to prove it.

There were no more than thirty-five people in the room, most of them from SCLC.

Dr. King was visibly angry. After looking about the room, he called to Ricks.

"Come here, son."

Ricks moved forward and stood in front of Dr. King.

"I've been out here fighting for a long time and I know what I'm doing. I'm in charge here and I intend to remain in charge. You can't hurt me. Remember that. You are not Martin Luther King! I'm Martin Luther King. No matter what you do, you'll never be a Martin Luther King."

Ricks did not reply. There was no need. For all practical purposes, the meeting was over. Nothing had been settled, but no one was surprised. We hadn't really expected to settle anything.

We left the room with the certain knowledge that there was a *wide* breach between them and us.

To keep the dissension from becoming public, SNCC's members made three key decisions: (1) we would continue to provide minimal support to the Selma campaign; (2) we would refrain from openly criticizing Dr. King and SCLC and (3) we would shift our base of operation from Selma to Montgomery.

We left a small group of SNCC people in Selma. They were instructed to remain in the background and to allow SCLC to make the decisions. The rest of us moved on to Montgomery. Still angry, we were intent on opening a "second front."

In Montgomery, we began working with a group of students from Tuskegee Institute. The students, many of whom had never participated in a demonstration before, wanted to march on the Capitol. They had sought direction from SCLC with no success. They agreed that something had to be done before the slowness of the courts destroyed the momentum that had resulted from the beatings on the bridge. They also agreed that the focus should be shifted from Selma to the entire state.

The small group of students marched on the state capitol on Wednesday, three days after the bridge beatings. As soon as they reached the Capitol Building, they were surrounded on three sides by city police and state troopers. When they were refused permission to take their petition into the Capitol Building, the students turned the march into a vigil. Plopping down on the cold, damp cement, they began to sing freedom songs. I stood about thirty feet away, outside the perimeter of the police lines, listening incredulously as chorus after chorus of "We Shall Overcome" sliced through the crisp tension in the air.

Rather than resort to overt violence, the police tried persuasion. When they realized that this wasn't going to work, they began to get tough. Moving in close, they informed the students that anyone who left the group would not be permitted to return. When we tried to deliver sandwiches and portable toilets to the students, the police turned us away.

A few of the students, the more modest ones, left the vigil in order to use the toilet. Those who remained were forced to stand and squat behind picket signs, to urinate in the street. This led to someone's calling the vigil a "piss-in."

Soon after the sun went down, it got very cold. Then it began to rain. By 2 A.M., the students were soaking wet. Many of them began to cough and sneeze. The situation soon became unbearable. After a discussion, the students, who were accompanied by Jim Forman, decided to move the vigil to First Baptist Church, a black church located about a block away. They spent the remainder of the night sleeping on the cold floor in the basement.

A number of very confusing things happened during the next couple of days: the students moved to Dexter Avenue Baptist Church, once pastored by Dr. King, and set up a vigil; police surrounded the church and refused to let anyone but SCLC's James Bevel enter; Bevel spent most of his time trying to get the students to dismantle their vigil and leave the church and SNCC continued to make hasty decisions, trying to make the best of a bad situation.

After much discussion of the student vigil, we decided that it could be used to draw attention away from Selma, where Dr. King was still waiting on the courts to give him permission to march. Although there was a tremendous amount of pressure on the students—the church's officials had turned off the water, heat and electricity—we were confident that they could hold out until we mobilized support. Stokely was inside with them. We were *certain* that he would intuit the importance of remaining in the church and, therefore, neutralize Bevel's arguments to the contrary.

Jim Forman jumped the gun. Working alone, he sent out a nationwide call for people to come to Montgomery and support the students in the church. The rest of us had been planning to organize support, but only on a local level. After Jim made his call, we considered it imprudent to contradict him.

On Friday afternoon, a huge truck pulled up in front of the building we were using as a makeshift headquarters. It was filled with camping equipment: tents, raincoats, boots, outdoor cooking utensils, plastic helmets and sleeping bags. It was more than enough to outfit one thousand people. I was astounded. The delivery men didn't know who had ordered the stuff.

Because our office was too small to store the unexpected supplies, they had to be placed on the sidewalk. Because no one was assigned to guard them, small children from the neighborhood began to walk away with those items which caught their fancy.

I got a second major surprise later in the evening while on my way to the church to see how the students were holding out. Although it was dark, the figure approaching me was unmistakable. It was Stokely. Rushing toward him, I began firing questions.

"What in the hell are you doing out here? Why in the hell aren't you in the church with the students? Where *are* the students? What in the hell is going on this goddamn town anyway? *Why did you leave the church?*"

I was so upset that I didn't immediately notice that Stokely was acting strangely. After a moment, when I realized that he wasn't responding to my questions, I calmed down and looked at him closely. There were large tears rolling down his cheeks. He didn't know where he was. The tension, frustration and fatigue had proven to be too much. He had freaked out completely.

I took him by the arm and headed back to our headquarters. I knew that he would be okay there after he had rested. The same thing had happened to several workers during the preceding summer. While we walked, I tried to get him to tell me why he and the students had left the church, but he was unable to do so. After leaving him at the office, I found some of the students who had been in the church with him. They said that the tension and isolation of the previous week had left them exhausted and unsure of what they were doing. They left the church because they had run out of reasons to remain.

Later that night, we decided to send Stokely on a speaking tour in the North. That would permit him to rest and get himself together. We also decided to cancel our plans for a Montgomery camp-in. There was no longer a focus for it. Moreover, it looked as if Dr. King was finally going to get permission from the courts to march.

When the big march finally did get under way at 12:47 P.M. on March 21, we were part of it. We joined because it was our only option. Jim had sent out a nationwide appeal; he'd spent over ten thousand dollars; several of the people who'd responded to that appeal had been brutally beaten in a Montgomery demonstration; John Lewis had been hospitalized as a result of the beating he had received on the Edmund Pettus Bridge. We were in too far to back out.

We did everything we could during the fifty-four-mile trek to be helpful. We still believed, however, that the march was a gigantic waste: in terms of money, human resources and human lives. At least three people were killed during the bloody, divisive campaign—Jimmie Lee Jackson, Mrs. Viola Liuzzo and the Reverend James Reeb.

I tried to share the magnificent enthusiasm of the huge throng that gathered on the grounds of the Alabama state capitol on the last day of the march. I didn't want to be bitter. I really wanted to believe the eloquent words that rolled from the mouth of Dr. King:

> My people, my people, listen! The battle is in our hands. . . . I know you are asking today, "How long will it take?" . . . I come to say to you this afternoon, however difficult the moment, however frustrating the hour, it will not be long, because "truth crushed to earth will rise again."
>
> How long? Not long, because "no lie can live forever!"
>
> How long? Not long, because "you shall reap what you sow!" . . .
>
> How long? Not long, because the arc of the moral universe is long, but it bends toward justice.

When it was all over, I headed back to Holly Springs—to pick up the pieces and to continue the struggle. Throughout the long, lonely drive, Dr. King's haunting "How long?" reverberated back and forth in my mind, between and around my solitary thoughts.

CHAPTER 10

Anarchists, Floaters and Hardliners

SNCC EMERGED FROM the Selma campaign and the long winter of 1965 beset by numerous problems. The zip and enthusiasm that had kept us going in previous times were gone. Although most of us were under twenty-five, we seemed to have aged. Our faces were haggard, our nerves overwrought. Arguments over trifles dominated all our gatherings.

Things would not have been so bad if the organization hadn't been going through a period of phenomenal growth. Between the spring of 1964 and the summer of 1965, SNCC, which had always been a small and close-knit organization, grew from 60 to more than 200 full-time staff members. During the same period, it added approximately 250 full-time volunteers. The annual budget grew from $20,000 a year to more than $800,000.

This growth, coupled with the changing nature of the struggle, was responsible for the emergence of several opposing factions. Although SNCC had always contained individuals who strongly disagreed with each other on various minor issues, it had never really had to contend with large factions divided by basic political differences.

I spent much of the spring and summer of 1965 attending long, involved staff meetings where the various factions haggled and argued over everything from the "true nature of freedom" to the cost of insurance for the organization's much abused fleet of automobiles.

The most flamboyant faction was composed of a group of "stars," who at various times were referred to as "philosophers, existentialists, anarchists, floaters and freedom-high niggers." Most of them were well educated. And they were about evenly divided between whites and blacks. They were integrationists who strongly believed that every individual had the right and responsibility to follow the dictates of his conscience, no matter what.

They were "high" on Freedom, against all forms of organization and regimentation. If a confrontation developed in Jackson, Mississippi, and a group of Freedom-High Floaters was working in southwest Georgia, they would pile into cars and head for Jackson. They might return to Georgia when the Jackson confrontation was over—and they might not. No one ever knew for certain what they were going to do or where they might turn up next.

They were great talkers, who generally ended up dominating those meetings and conferences they saw fit to attend. Holding forth with long, involved existential arguments, they would take as long as three days of nonstop talking to win a single inconsequential point; but they didn't mind.

They loved to bring meetings to a screeching halt with open-ended, theoretical questions. In the midst of a crucial strategy session on the problems of community leaders in rural areas, one of them might get the floor and begin to hold forth on the true meaning of the word *leader.*

"Well, I think we have to examine our premises. What is a leader? It seems to me that we can't go on until we examine and define this term. Why do people choose to become leaders? SNCC people say they want to develop grass roots leaders. Maybe this is wrong? Is it really possible to 'develop' leaders? Maybe what some people say is true. Maybe leaders are born and not made, not developed.

"Maybe we aren't helping these people at all. Maybe we're just substituting one form of oppression and manipulation with another. We say we are trying to help people to acquire freedom, you know. Well, what *is* freedom?"

My characterization of the Freedom-High position probably reveals my bias. I was one of their staunchest opponents. I considered them impractical. SNCC was not a debating society. It was an action organization.

Those who supported the faction to which I belonged were known as "Hardliners." They were primarily black. We were moving in a Black Nationalist direction.

Before 1965, no one had ever been forced out of SNCC. This was true even though some members had grossly abused the organization's rules and regulations. We Hardliners were committed to the position that everyone had to accept at least a minimal amount of discipline. We repeatedly argued that those who were too "free" to accept and abide by the organization's rules should be fired.

The Hardliner-Floater schism probably wouldn't have become important had it not been for three things: (1) the attacks being made against SNCC by various government officials; (2) the passage of civil rights legislation by the federal government, especially the Voting Rights Act, and (3) our continuing confusion about who we were and what we should be doing. All these things contributed to a great deal of tension, frustration, paranoia and internal hostility.

The attacks by government officials, which weren't nearly as vicious as they would become, bothered us a lot. We were being red-baited. There were constant rumors that the House Un-American Activities Committee was going to investigate us. Police officials in various sections of the nation, Philadelphia in particular, were harassing our members. Foundations that had offered us small but important support in the past, began to ignore our letters and leave our calls unanswered. We sensed this change of mood and it troubled us. More and more, people were beginning to treat us as if we were outlaws.

The civil rights bills passed by the federal government had the effect of undermining SNCC's basic thrust. As long as blacks were being denied the use of public accommodations, the demand for black access in this area was revolutionary. As long as poor blacks in the rural South were being denied the basic right of every citizen in a truly democratic society, voter registration was revolutionary. When the federal government passed bills that supposedly supported black voting rights and outlawed public segregation, SNCC lost the initiative in these areas.

This was true even though the civil rights bills did not begin to deal with the basic problems of blacks. They had no effect on the South's rigid system of caste and class oppression. They were not addressed to the economic oppression of poor blacks. All they really did was create an *illusion* of progress. Our task, therefore, was to develop new programs designed to widen the struggle and expose the continuing contradictions between America's bright promises and dreary realities.

The identity problems that plagued SNCC's members from the organization's origin were greatly magnified during this period. We all realized that SNCC was no longer a student organization. Over the years, the organization had moved far away from the comparatively tranquil Negro college campuses of its origin. Moreover, almost none of the organization's members were students. We were, despite our youth, professional revolutionaries, searching for the pulse of the revolution.

Two questions obsessed us: What kind of revolutionaries are we? What kind of struggle should we launch?

The opposing positions taken by the Floaters and Hardliners grew out of differing attempts to provide comprehensive answers to these questions. Both groups were convinced that they were right. Had it not been for the confusion that dominated the atmosphere, the Floaters probably would have become dominant. They were closer to SNCC's free-wheeling, anarchistic origins. The only thing that kept us Hardliners in the running was our "guarantee" that we could move SNCC beyond the morass in which it was stuck.

Of the numerous confrontations between the Floaters and Hardliners, I remember one in particular. We were in Atlanta for our annual fall meeting at Old Gammon Seminary. I was very tired and preparing for bed one night when Stanley Wise, a former NAG member who had joined SNCC, rushed through the door and told me that the Floaters were conducting a rump meeting.

"Where are they?" I asked as I reached for my pants.

"They're over at Judy Richardson's apartment!"

Stanley, who was very excited, explained that the meeting had been going on for almost an hour.

"I was getting them for a while, but they had me outnumbered."

While I searched for my shoes through the dirty clothes and paper that cluttered the floor, a tall, light-skinned fellow entered the door. He sported an unusual moustache, a wispy affair that curled over his top lip. He said that his name was Hubert—Hubert Brown. "My friends call me Rap," he added. He was looking for Fred Mangram, my roommate.

"He's not here, but he may be where we're going," I told him.

Normally we did not allow nonmembers to attend SNCC gatherings, but I knew Rap was okay. His brother Ed was an old SNCC veteran and a good friend of mine.

"If you're as good a man as Ed, we'll need you," Stanley replied to Rap as we headed toward the car.

There were about fifteen people sitting around Judy's sparsely furnished living room when we arrived. A hush settled over the room when the three of us entered.

The lines had already been drawn. They knew who we were and where we stood. Because they outnumbered us, we split up. That was to keep them from getting a clear focus. I stood on one side of the room and Rap stood on the other. Stanley moved to the corner occupied by Moses.

Casey Hayden, who had been speaking when we entered, continued after a moment.

"Do you remember when you were a child? Do you remember how people oppressed you, not with chains or anything, but because they were always trying to get you to do things you didn't really want to do?"

"What has that got to do with SNCC and the work before us?" I fired at her.

While she was answering, Stokely arrived. He took a seat on the floor and listened.

As I had expected, we got into a big argument. The three of us—me, Stanley and Rap—were doing very well for a while. We wouldn't allow them to move into the abstract, philosophical areas where they were most comfortable.

"Stick with reality," I kept yelling. "We are trying to move people from one place to another. Sometimes we have to coerce them. Sometimes we have to shame them. They're frequently afraid and reluctant to do the things we want, but that's the way it is. We are not oppressors. We aren't doing anything we should be ashamed of. We *have* to establish priorities. Getting people to deal with their fears and insecurities is a SNCC priority. There's nothing wrong with that! We don't need to get hung up on a lot of philosophy. What we ought to be discussing is strategy and programs. Where are your programs?"

This line of reasoning was working very well until Stanley got caught in a trick bag. He had been working in Cambridge, Maryland, where the CNAC had suffered several setbacks. Stanley had some authority in the CNAC and had taken the setbacks very hard. He was very sensitive about them.

"Stanley! Are you going to stand there and say that you have *no* reservations about what happened in Cambridge? Are you going to stand there and say you did not manipulate people in Cambridge? Don't you really believe that you failed in Cambridge? What did you really do to help those people?"

It was a personal attack. Not unfair, however. SNCC people considered personal feelings and insecurities to be legitimate topics for debate and argument. We did not accept the normally recognized dichotomy between personal and public concerns. To us, *everything* was political.

Stanley tried to defend himself, but they were hitting too close to home. He broke down and started crying. They knew they had him then. Their questions and comments grew sharper. After an hour, he couldn't take it anymore. He left the meeting in tears.

Stokely, who had been quiet through the entire affair, spoke up shortly after Stanley left. He asked me if I were absolutely certain that we were justified in *everything* we'd done. He recalled several campaigns we had been involved in: Cambridge, Holly Springs, Montgomery, the March on Washington, Selma. He wasn't accusing me. He was trying to come to grips in his own mind with our actions, trying to decide how he really felt about what we had been doing.

"Cleve, are you absolutely certain that you have *never* manipulated anyone during that whole time?" he asked.

Before I could respond, Ivanhoe, who had been silent, spoke up. Although he wasn't aligned with the Hardliners, he was not a Floater. His comments were typically convincing.

"I think you're missing the point, Stokely. Cleve and Rap aren't saying that they haven't manipulated anyone. That isn't the issue. The issue for any revolutionary is not essentially related to the question of manipulation. Manipulation is irrelevant. We must be concerned with goals and purpose.

"The MFDP is a good example. We went to Atlantic City with legitimate goals. Some people got caught in the middle and were destroyed—Chaney, Goodman and Schwerner are obvious examples. We haven't lied to anyone. This is dangerous work. People have gotten killed. They are going to continue to be killed. No one can guarantee a mistake-free life, certainly not a mistake-free revolution. We have to keep on getting up—doing the best we can, learning from our mistakes and not worrying about the rest!"

The three of us—me, Rap and Ivanhoe—took them to the cleaners from that point. It was five in the morning before we were finished. A very important thing happened during the course of the night, however. Bob Moses withdrew from the discussion. He didn't withdraw because we were winning or because he was tired. He did so because he realized that those aligned with him, the Floaters, were incapable of pursuing his ideas as well as he.

There is no doubt in my mind; Bob could have won the confrontation if he had wanted to. He was smart enough to tie all three of us into philosophical knots. He didn't do it because he realized that he would have had to become what he abhorred, a manipulator, in order to do so.

I have always believed that Bob Moses began a gradual withdrawal from SNCC that night. Shortly thereafter, he dropped his last name and sought to become a different person, Robert Parris. Parris was his middle name. I was there on the night when he changed his name. It was one of the most remarkable ceremonies I have ever witnessed.

It was about nine in the evening and we were gathered in one of the cavernous meeting rooms at Old Gammon. There were about seventy-five or eighty people present, including Bob's wife, Donna. The meeting, which had been going on for a couple of hours, was about to break up. Nothing important had happened. Just before the chairman called for adjournment, Bob stepped to the front of the room.

"I have a message for you," he announced. Everyone was immediately quiet. "I have changed my name. I will no longer be known as Robert Moses. From now on it will be Robert Parris. My wife's name is no longer Donna Moses. She will be known from now on as Donna Richards.* We are still married, but we are changing our names."

Before anyone had an opportunity to respond, he launched into a complicated existential exposition. I couldn't understand all of it. It was too heavy. He said that he was changing his name because he had been under a lot of pressure in Mississippi, that a lot of people and things were pulling at him and he wanted to get

---

* Her maiden name.

away. He said that changing his name was the only way to cope with the situation.

He said that the strain he had been under was not positive. Then, he began to talk about his childhood and his mother.

"She got sick once because we were poor and there was a lot of strain and tension in our home. We called the police and they sent an ambulance. The ambulance took her to Bellevue Hospital. I was very young then. My father took me with him when he went to see her. He asked the doctors what was wrong with her and they said she was crazy.

"My father became very angry and started to scream at the doctors. 'She's not crazy! She's not crazy! She's not crazy! You're the ones who're crazy!'"

Although Bob may have been on the verge of hysteria, he appeared to be in complete control of himself. He told us that he was drunk, but he obviously wasn't. Moving back and forth across the front of the room, he began to talk about his father.

"My father was a poor man. He always worked very hard. Sometimes he would get confused about who he really was. His name was Gary and he sometimes thought he was Gary Cooper, the white movie star. He loved us all very much."

His language was very delicate, almost like poetry. We were quiet and very still. It was as if we were watching a tremendous ceremony, a sacred performance. When he was finished talking about his father, he bowed. One hand in the front, tucked in at the waist, and the other at his rear.

We thought he was finished. Some of those present began to clap. They stopped abruptly when he produced a bottle of wine and a large cut of cheese.

"I want you to eat and drink," he said. He explained that it would be a part of the symbolism. He moved forward and passed the bottle and cheese. I was startled when the bottle was passed to me. Although several of those who had handled it before me had placed it to their lips as if they were drinking, it was empty. The inside of the bottle was dry. We all ate the cheese.

After the bottle was passed to everyone in the room, it was returned to Bob. Donna moved forward and joined him and before anyone could speak, they left. The room was quiet, very quiet.

"Did he mean it?" someone whispered to me as I moved out of the room toward the darkness outside.

"Bob means everything he says," I replied.

In November, 1965, SNCC held its annual election. In previous years no one had paid very much attention to them. John Lewis and Jim Forman were generally reelected with no opposition and little discussion. Because of the differences between the Hardliners and Floaters and the positions John had taken on numerous controversial issues, the elections attracted more than the usual amount of attention.

In all the meetings before the actual voting, the Floaters charged that SNCC was becoming too bureaucratic. They argued that the office workers, especially those in Atlanta, were exerting too much authority over those who worked in the field. They wanted less bureaucracy and more decision-making power for the field workers.

For once, the Hardliners and Floaters agreed. We Hardliners agreed that the organization's decision-making apparatus was in need of change. Of course, our reasons for supporting this position were different from those of the Floaters. We felt that the existing structure was too inefficient. We wanted the organization restructured so that new programs could be organized and implemented.

We also believed that John Lewis should be replaced as SNCC's chairman. We cited his actions in Selma and called attention to the thousands of dollars the organization had wasted as a result of his stubbornness. We also suggested that his ideas and opinions no longer reflected those of the majority of SNCC's members. Almost everyone, including the Floaters, agreed with this position. However, there was little agreement about what should be done.

There was some dissatisfaction with Jim Forman, too. This did not follow factional lines. Several members were concerned about the camping equipment Jim had purchased in Montgomery. They argued that no individual should have the power to spend such a large amount of money without the consent of others. The only thing that kept Jim from coming under a great deal more fire was the fact that at the time no one was really positive that he was the one who had ordered the camping supplies.

In any event, a compromise was reached. We agreed to restructure the organization and place ultimate decision-making power in the hands of the Executive Committee. The Executive Committee, which was to have twenty members, would make important decisions during those periods when the entire staff could not be consulted. Those elected to the Committee represented every faction of the organization.

Jim Forman came up with the idea that proved to be the most crucial to the balance of power in the organization. He suggested that a Secretariat be set up to make day-to-day decisions. The three members of the Secretariat would be the executive secretary, the chairman and the program chairman. This suggestion received wide support and was adopted without significant opposition.

The election of the Secretariat provided me with one of the biggest surprises, and certainly one of the biggest honors, of my twenty-one years. Jim Forman was reelected as the executive secretary. John Lewis was reelected chairman. And I was elected program secretary.

CHAPTER 11

Setting the House in Order: Seeking Levers of Power

AFTER THE ELECTION, we conducted a series of crucial meetings. They were devoted to setting SNCC's house in order. The organization and each of its members were thoroughly evaluated. All those programs adjudged unproductive were summarily discontinued.

Letters were sent to every member who did not have a specific job. Recipients of the letters were instructed to inform me or the Executive Committee of their activities. A second letter, sent to those who did not reply, informed them that their checks were being discontinued and that they could consider themselves ex-members of the Student Nonviolent Coordinating Committee.

The letters generated a great deal of resentment. SNCC had never operated in such a manner before and a lot of people assumed that I was responsible, since the letters bore my signature. They were only partially correct. Although I signed the letters, they had the unanimous endorsement of the Executive Committee.

The Floaters were particularly incensed. Several of them called the National Office and cursed me, others sent insulting replies. Some vowed they would never comply with the new directives coming from my office.

In the late fall things came to a head at a general staff meeting at a small resort in Waveland, Mississippi. I was in charge of planning the meeting. Because I knew that there would be a lot of tension, I took special pains to ensure that everyone would be as comfortable as possible. I was determined to make it the best organized, best run staff meeting SNCC had ever had.

I devised an elaborate schedule. A registration desk was set up for those who arrived at night. All they had to do was sign in, whereupon they would be assigned meal tickets and sleeping quarters. The best bedrooms were reserved for married couples. I even had printed programs with information on everything from the subjects to be discussed in the workshops to recreation periods.

"I don't want them to have anything to complain about," I told Cynthia Washington, who was helping me.

Of course, nothing went according to plan. On the first night, a group of Floaters got together on the lawn outside the registration headquarters and burned their meal tickets. They made a big ceremony out of it. Laughing and joking, they made plans to hold a caucus the next morning to restructure the agenda for the meeting. Most of them refused to register.

Jim Forman and I stood listening and watching while they cursed the "regimentation" that we were ostensibly trying to force upon them. Jim was nervous. I was angry. The confrontation that everyone had been trying to avoid for almost six months was at hand. Late that night, I got a crew together and made plans for the following day.

The Floaters did not get up for breakfast. They had stayed up too late the night before. When they arrived at the cafeteria at lunch time, we had a delegation on hand to meet them. Ed Brown, Roy Shields, Jimmy Jones and several others were stationed inside the cafeteria. I was at the door.

"No one eats without a meal ticket," I informed them.

It was a stalemate. They hadn't expected us to make a stand in the cafeteria. They retreated. When they returned a few minutes later, they were armed with pool cues, baseball bats, knives and a couple of pistols. Cynthia saw them coming and ran to get Jim Forman. We were on the verge of coming to blows when Jim arrived.

"Why don't you ease up?" he asked me.

"No!" I replied. "These guys came here to do just what they're doing. They've got it so nobody in the organization can do a goddamn thing. Every time we try to plan something, they fuck it up with a lot of irrelevant bullshit. Time's up for that kinda shit."

We were all inside the cafeteria at this point. Jim was between us, looking very sad. Cynthia was standing at the entrance of the kitchen, prepared to punch meal tickets. While the two groups argued back and forth, Donna Richards moved toward the entrance of the kitchen. Because she didn't have a meal ticket, Cynthia reached out to stop her. But Donna was too fast. She darted past Cynthia and leaped into the kitchen.

As soon as she stepped through the door, one of the cooks, who was apparently frightened by the knives, bats and pool cues, took a swipe at her with a huge meat cleaver. The cook, who was a member of the resort staff, must have thought that Donna was trying to attack her. Only luck kept Donna from receiving a horrible injury.

The incident led to bedlam. Courtland Cox, a former NAG

member and a big Floater leader, jumped on a table in the middle of the room and tried to talk over the shouting and screaming. He was yelling at the top of his voice, but I couldn't understand a word he was saying even though I was standing nearby. Ruth Howard, another Floater, was also screaming. It was obvious that she was hysterical. John Lewis rushed forward and slapped her hard on the cheek. She started to cry.

Faced with pandemonium threatening to become war, Jim left the room. Everyone saw him and understood that by doing so he was renouncing responsibility for what happened from that point on.

Fortunately, Forman's departure motivated several uncommitted bystanders to come over to our side. Looking about them, the Floaters realized that we outnumbered them four to one. They backed down and left the cafeteria. From that moment on, it was clear that the Hardliners were in control of SNCC. Everyone who ate in the cafeteria for the remainder of the staff meeting used a meal ticket.

After Waveland, my mandate was clear: shake out the deadwood and get the organization moving. That's just what I attempted to do. For instance, there was a black SNCC volunteer in Jackson, Mississippi, telling high school kids to drop out of school because it was irrelevant. She had a degree from an exclusive white women's college in the North. When the head of the project in Jackson tried to get her to stop, she ignored him. I took a trip to Jackson.

I told her what we were trying to accomplish in Jackson and how she could help us. Her answer wasn't good enough. Soon afterward she was an ex-SNCC volunteer.

Relations between me and Stokely were very bad during this period. As a matter of fact, we weren't even speaking. When we had something to say to each other, we used Jim Forman as a go-between. Our differences resulted from my actions as program secretary. Stokely thought I was more concerned with cars, money and equipment than with people. He was convinced that I had become a bureaucrat.

Many SNCC members resigned during this period. Most of the resignations came from those who realized that they were out of step with the organization's new cadence.

Robert Parris was one of those who resigned. His reasons were different from most, however. He explained in his letter that he was resigning for the same reasons he had changed his name: the people with whom he was working depended on him too much. He said that his move from Mississippi to Alabama had not produced the desired result. He was too well known.

By the end of 1965, SNCC was moving in several directions. Although we had pretty much eliminated the Hardliner-Floater schism, we were still trying to establish a coherent identity and realistic programs. Every meeting, whether large or small, devoted some portion of its time to two crucial questions: Who are we? What should we be doing? I now understand some of the reasons we had so much trouble answering them.

By the end of 1965, SNCC had gone through two distinct phases and was entering a third. During the first phase, which lasted from 1960 through 1962, SNCC's members were largely middle class and very religious. They were primarily interested in securing integrated public accommodations and basic dignity for blacks. Depending on courtesy and a Gandhian ability to take punishment, they were basically successful.

SNCC entered its second phase when it sent organizers to Mississippi to register rural blacks. Many of those who joined the organization during this period, which roughly lasted from

1962 through the Democratic National Convention in August, 1964, saw themselves as guerrilla organizers. I was one of them. The major difference between us and those who had joined the organization earlier was that we were never committed to nonviolence as anything more than a tactic. Although we, too, were politically immature, we were not as rigid as those who joined SNCC before us.

Even though we believed that we were morally superior to those aligned against us, we did not believe that we were fighting to create a "morally straight" America. We were essentially concerned with power. Integration was never seen as anything more than a means to an end. We were convinced that we could precipitate a redistribution of wealth and power. Our failure at the Democratic National Convention taught us that the struggle was going to be longer and far more difficult than we had initially realized.

As I have indicated, SNCC entered a third stage after Atlantic City. That stage continued at least until the spring of 1966, when Stokely was elected chairman.

In the meantime, SNCC's members were at work in almost every section of the nation, searching for levers that might be used to spring blacks and poor whites into power. We were all very conscious of the fact that the axis of the struggle appeared to be shifting away from the rural South to cities in the North. The totally unexpected rebellions in Harlem, Watts, Chicago and Philadelphia made a big impact on our thinking. They motivated us to begin a search for ways in which we could mold the discontent in the urban ghettos to revolutionary advantage.

Our search took place within the framework of key questions: What is the relationship between corporate affluence and urban poverty? What is the ideal form of government for a city? What are the best issues around which people in cities can be organized?

What is the value of the vote in cities? Should we attempt to develop alternative or parallel governing structures in cities as we have in the rural South?

Ivanhoe, who had left the South and settled in Columbus, Ohio, shortly after the conclusion of the Selma campaign, presented a position paper at a midwinter staff meeting that dealt with many of these questions. In the paper, which was well received, he explained that he had been busy organizing a "community foundation" that SNCC might consider as a model for neighborhood self-government in urban areas.

The neighborhood in which he was working covered an area thirty blocks wide and fourteen blocks long. Seven thousand people lived there, many of them unemployed. These people, who were black and white, elected a governing board for the community foundation. The board was given the authority to make decisions about how money would be spent in the community. It was also given the power to make grants to local people from a community treasury. The grant money came from local fund-raising projects.

The foundation, which gave the vote to all residents in the neighborhood over sixteen, was, in effect, an incorporated community pursuing real "participatory democracy."

Sometime during the winter of 1965, Julian Bond, who had been serving almost five years as SNCC's communications director, poet laureate and man Friday, decided to run for a seat in the Georgia House of Representatives. He made the decision after a Supreme Court reapportionment decision resulted in the creation of an all-black district in a desperately poor section of Atlanta called Vine City.

Some SNCC members disagreed with his decision. Their opposition grew out of their belief that electoral politics were thoroughly corrupt and would, in some way, corrupt Julian and SNCC. They were in a minority, however; SNCC agreed not only to endorse and support Julian, but also to lend him the five hundred dollars he needed to file for office.

Although I was one of those who supported Julian's candidacy, I must admit that I had no clear idea of what he would be able to accomplish if he were elected. His preliminary support of a two-dollar minimum wage law, abolition of the death penalty and the removal of all voting requirements except age and residence sounded good to me. I also agreed with his plan to conduct "people's conferences" in order to find out what his constituents wanted.

In addition to our concern for Julian's campaign, Stokely's work in Lowndes County and Ivanhoe's project in Columbus, we worried about the war in Vietnam. We spent a great deal of time trying to develop an analysis of the relationship between the war and our struggle. Several members were pushing for an official SNCC statement condemning the war. Others cautioned against such an act. They called attention to the fact that no other black protest organization had come out against the war. They argued: "We will almost certainly be singled out for increased repression if we take a stand against the war before anyone else."

This was not a popular position. The majority of the organization's members were thoroughly convinced that we had no alternative but to condemn the war and the American government. During one staff meeting when a discussion of the war became particularly heated, Courtland took the floor and waxed eloquently.

"I want you to think with me, to see some parallels," he said. "Mississippi and Vietnam; they are very much alike. Think about Vietnam's Ky and Senator James O. Eastland; they are very much alike. Think about the problems of Mississippi's poor, disenfranchised blacks and the problems of Vietnam's poor, disenfranchised peasants; big business is making a killing in both places. Also, consider the similarities between Vietnam's National Liberation Front and SNCC. They ought to be very much alike!"

On January 3, 1966, less than a month after Courtland made these remarks, Samuel Younge, Jr., was shot to death by an irate filling station attendant in Tuskegee, Alabama. Sammy, who was a former navy man, was attempting to use the "White Only" restroom at the filling station when he was killed. His death was a blow to us because he was a key figure in student demonstrations at Tuskegee Institute.

The absolute absurdity of a man having to die for attempting to do something as basically human as using a toilet filled us with rage. Like the thousands of students who demonstrated in cities around the nation in protest of his death, SNCC's members could not help recognizing the gross contradictions between the freedom that Americans were killing and dying for in Vietnam and the race hatred that motivated Sammy's murderer. The fact that Sammy was a veteran made his death all the more galling.

On January 6, 1966, SNCC issued a statement on the war in Vietnam:

> The Student Nonviolent Coordinating Committee has a right and a responsibility to dissent with the United States foreign policy on any issue when it sees fit. The Student Nonviolent Coordinating Committee now states its opposition to United States' involvement in Vietnam on these grounds:
>
> We believe the United States government has been deceptive in its claims of concern for the freedom of

> the Vietnamese people, just as the government has been deceptive in claiming concern for the freedom of colored people in such other countries as the Dominican Republic, the Congo, South Africa, Rhodesia, and in the United States itself. . . .
>
> The murder of Samuel Young [*sic*] in Tuskegee, Alabama, is no different than the murder of peasants in Vietnam, for both Young [*sic*] and the Vietnamese sought, and are seeking, to secure the rights guaranteed them by law. In each case the United States government bears a great part of the responsibility for these deaths.

SNCC was inundated with a storm of criticism as soon as the statement hit the streets. Congressmen, senators, right-wing organizations and leaders of other civil rights organizations all said the same thing: the war was not our concern and we had no business whatsoever criticizing it or the government. Several major contributors notified us by telegram and telephone that they would no longer support the organization.

Julian Bond caught a great deal of the flack generated by the statement. When a reporter called him on the telephone and asked if he supported the statement, Julian said yes. The reporter, who was illegally taping his conversation with Julian, subsequently released the tape. This resulted in Julian's being denied his seat in the Georgia state legislature.

While the nation's attention was focused on the Georgia legislature and Julian's attempt to gain his seat, something very exciting was happening in Lowndes County. Although the Lowndes effort began in much the same manner as virtually every other

SNCC project, it was different in one very important aspect: blacks constituted 80 percent of the county's residents.

Lowndes County stood out for other reasons, too. Whites controlled *everything*. Before March, 1965, when SNCC workers first went to work in the county, none of its black residents was a registered voter. This was true even though 130 percent of the whites were registered. Some 60 percent of the county's blacks were farmers, most of them share-croppers. Half of the black women were employed in white homes as domestics, and the median income for blacks was $935 a year. The median income for whites was almost five times higher. Eighty-six white families owned 90 percent of all land in the county and their relatives controlled all elected and appointed offices.

Our plan for Lowndes was simple. We intended to register as many blacks as we could, all of them if possible, and take over the county. An obscure Alabama law that made it relatively simple to start a new party gave us the edge we needed. We believed that a complete victory was possible. After achieving success in Lowndes, we intended to widen our base by branching out and doing the same thing in surrounding counties. We were convinced that we had found The Lever we had been searching for.

As spring moved into summer and summer into fall, interest in the Lowndes County struggle began to grow. Stokely, who was working around the clock in order to get ready for the November election, began to attract supporters. He was proving to be a dynamic leader and his influence within the organization began to increase.

We were also thinking black. As a matter of fact, Lowndes was the first SNCC project in which the emphasis was completely black. There was no talk about an integrated ticket for the party, which was officially known as the Lowndes County Freedom Organization (LCFO). The party's symbol, a snarling black panther, was self-explanatory.

Several outsiders accused us of being black racists. That wasn't true. We weren't racists. We were just convinced that it was time for blacks to begin to work by themselves; to prove, once and for all, that blacks could handle black political affairs without assistance from whites.

Our candidates were running for seven positions: sheriff, coroner, tax assessor, tax collector and three seats on the Board of Education. They were all very strong local people who understood that they were risking their lives* by running for office in a county in which no black man had voted for seventy-five years.

On Monday, November 7, 1966, the night before the big election, we had a huge rally in Mount Moriah Baptist Church near Hayneville, the county seat. The 650 people who crowded into the church to gird up their confidence and to receive final instructions for the big day were aware without being told that the

---

* On August 20, 1965, the Reverend Jonathan Myrick Daniels, a twenty-six-year-old seminary student from the Episcopal Theological School in Cambridge, Massachusetts, was shotgunned to death in Hayneville, Alabama. Richard R. Morrisroe, a white Roman Catholic priest who accompanied him, was seriously wounded by a shotgun blast during the same attack. The men, who had been in Lowndes County registering voters a few days before, were killed by Thomas L. Coleman, a part-time deputy sheriff. He shot them as they approached a small grocery store with two black girls, Ruby Scales and Joyce Bailey. The four of them were going into the store to buy food when they were confronted by Coleman with a shotgun. Mr. Daniels was killed while pushing Miss Scales out of the path of Coleman's fire and Father Morrisroe was shot while pulling Miss Bailey out of danger. Coleman was released from jail on a $12,500 bond the day after the shooting. Alabama Attorney General Richmond Flowers was ousted from the case when he tried to get it moved from Hayneville because he felt it was impossible for the state to get a fair hearing there. He was also unsuccessful in his attempt to get an indictment for murder.

next day, whatever its outcome, would mark the beginning of a new epoch.

There were several speakers. Some gave long speeches and others were succinct. Mrs. Alice Moore, the Lowndes County Freedom Organization's candidate for tax assessor, was one of them. She took the microphone, uttered one sentence, "My platform is tax the rich and feed the poor," and sat down to thunderous applause.

Stokely, who had been arrested in Selma the Friday before on a special warrant issued by the mayor, delivered a speech that was repeatedly interrupted by cheers and applause. The people loved and respected him. He was, as far as they were concerned, one of them. He wasn't their leader. He wasn't some outsider from the North, trying to tell them what to do and how to do it. He was a *member* of the community. His speech, which ranks with the best I have heard, was a combination of New Politics, old-time religion and "you-can-make-it-if-you-try" idealism:

> . . . We have worked so hard for the right to come together and organize. We have been beaten, killed and forced out of our houses.
>
> But, tonight says that we are right!
>
> We have done what they said we could not do. Colored people have come together tonight! Tonight says that we CAN come together and we can rock this country from California to New York City!
>
> When we pull that lever we pull it for all the blood of Negroes that the whites have spilled. We will pull that lever to stop the beating of Negroes by whites. We will pull that lever for all the black people who have been killed. We are going to resurrect them tomorrow.

> We will pull that lever so that our children will never go through what we have gone through. We don't need education—all we need is the will, the courage and the love in our hearts.
>
> . . . WE CAN STICK TOGETHER!
>
> . . . There are some who are not with us. When Moses crossed the Red Sea, he left some people behind. We are going to leave some Uncle Toms behind.
>
> We have a lot to remember when we pull that lever. We remember when we paid ten dollars for a school book for our children. We remember all the dust we ate.
>
> We are pulling the lever to stop that!
>
> . . . We say to those who don't remember—You better remember, because if you don't MOVE ON OVER, WE ARE GOING TO MOVE ON OVER YOU!

We filed out of the church into the dark, clear night *knowing* we were going to win. When we got up the next morning, we weren't quite as certain. As reports from the various precincts trickled in during the day, we grew less and less certain. The whites had launched an all-out effort to defeat our candidates.

They stuffed some ballot boxes; they forced some blacks to use ballots that had already been marked for white candidates; they insisted on helping those blacks who couldn't read; they brought in truckloads of blacks who worked on their farms and told them who to vote for; they refused to allow several of our watchers to work at the polls.

The polls closed at 6 P.M. The final returns confirmed what we already knew. Every one of our candidates was defeated. There were no tears or angry denunciations. We had fought the good fight and had nothing to be ashamed of. The spirit of the local people remained strong, unbroken. Despite their disappointment, they all understood what Mr. John Hulett, the LCFO chairman, had meant during his speech the night before when he anticipated the election's outcome:

"Our candidates represent the residents of Lowndes County. They also represent all the poor people in the country. No matter what happens tomorrow night, I will hold my head as high as I have ever done. It is a victory to get the black panther on the ballot."

CHAPTER 12

From Black Consciousness to Black Power

SNCC CONDUCTED THE most important staff meeting in its history during the second week in May, 1966. The meeting, which was held at a beautiful resort camp near Nashville, Tennessee, was dominated by good feelings and a renewed sense of hope. Everyone present seemed to believe that the tactics used to organize the Lowndes County Freedom Organization were what we had been searching for since Atlantic City. We were relaxed and confident. Those hours not spent in workshops and general meetings were devoted to swimming, spirited volleyball games and leisurely walks in the wooded hills surrounding the camp.

There was widespread agreement that it was time for SNCC to begin building independent, black political organizations. We were convinced that such organizations working together could end racial oppression once and for all. Most of our time was spent discussing the best ways to publicize and win support for this new approach to the struggle.

We were absolutely convinced that there was no viable future for blacks, poor blacks especially, within the Republican and Democratic parties. We were also convinced that the federal government, where we had looked for support in the past, was part of the problem.

A small minority of those present were worried about SNCC's new commitment to independent, black politics. "How can we create an integrated society if we are building racially segregated political parties?" they asked. This was a legitimate question. Most of us were convinced, however, that it missed the point.

"Blacks are not being lynched and dumped into muddy rivers across the South because they aren't 'integrated,'" we countered. "Black babies are not dying of malnutrition because their parents do not own homes in white communities. Black men and women are not being forced to pick cotton for three dollars a day because of segregation. 'Integration' has little or no effect on such problems.

"Look at all those 'integrated' towns and cities in the Midwest. Niggers up there have it just as bad as we down here. The real issue is power; the power to control the significant events which affect our lives. If we have power, we can keep people from fucking over us. When we are powerless, we have about as much control over our destinies as a piece of dog shit."

It was obvious that many of those whom we hoped to influence were not thinking in terms of power. We believed they lacked the proper point of view; they did not possess what we called "Black Consciousness." Throughout the staff meeting, we talked about Black Consciousness. It was new and exciting. More important, it helped us develop a better understanding of many conceptual problems that had stymied us since Atlantic City.

What is Black Consciousness? More than anything else, it is

an attitude, a way of seeing the world. Those of us who possessed it were involved in a perpetual search for racial meanings. Black Consciousness, which was an admitted consequence of the failure of the movement up to that point, forced us to begin the construction of a new, black value system. A value system geared to the unique cultural and political experience of blacks in this country.

Black Consciousness signaled the end of the use of the word *Negro* by SNCC's members. Black Consciousness permitted us to relate our struggle to the one being waged by Third World revolutionaries in Africa, Asia and Latin America. It helped us understand the imperialistic aspects of domestic racism. It helped us understand that the problems of this nation's oppressed minorities will not be solved without revolution.

From an organizational point of view, Black Consciousness presented SNCC with one major problem. More than 25 percent of our members were white. It was obvious that they did not and probably could not possess Black Consciousness.

A major confrontation was averted by the growing belief that whites should no longer be used as organizers in black communities. The consensus was that SNCC's white members should begin organizing poor whites. It was obvious that poor whites, who form the nation's largest oppressed group, would have to support any revolutionary movement that hoped to be successful. We believed that white organizers working with poor whites might be able—some time in the future—to create a second front to augment the efforts of those blacks struggling to change the system.

The workshops devoted to these matters were sometimes painful. No one wanted to hurt anyone else's feelings. Although it had always been an issue in the organization, the role of whites had never really been openly discussed. More important, no one had ever seriously suggested that jobs be assigned on the basis of race instead of competence. I witnessed several members leave the discussions with furrowed brows and tears welling in their eyes.

The most important portion of the staff meeting was the annual election. It began at about seven on the evening of May 14. There were three offices open: chairman, executive secretary and program secretary. Ralph Featherstone ran against me for program secretary. Because Jim Forman had decided not to run for another term, Ruby Doris was nominated for executive secretary. John Lewis, who had been chairman for several years, was opposed by Stokely and several others.

It began very informally. None of us had ever put much stock in official positions and titles. With the exception of Stokely's attempt to become chairman, the election was just another bothersome formality. For this reason, most of the three hours before the first vote were consumed by a wide-ranging philosophical discussion of programs. When the vote was finally taken, Ruby Doris was elected executive secretary, John Lewis was picked again as chairman and I continued as program secretary.

There was some dissatisfaction with this outcome. Someone, it may have been Jim Forman, was upset because only a few of those present had voted. While we were discussing the situation, Worth Long, an ex-SNCC member, entered the room. When informed that John had been reelected, he hit the ceiling.

Worth claimed that John had been in office too long and that his ideas no longer represented those of the majority of the organization's members. Instant chaos. Worth had hit a nerve, one much more sensitive than any of us had previously realized. Stokely had attempted without much success to win support for his candidacy during the weeks before the staff meeting. He now took heart.

Jack Minnis, a master political strategist, called for a caucus in one of the rear corners of the room. Stokely, Ed Brown, Roy Shields, Featherstone and I joined him. I knew what was on his mind before he said a word. We *could* get Stokely elected if we were shrewd enough. After talking for a few hurried minutes, we resumed our seats. I waited a couple of minutes before asking for the floor.

"There seems to be some question in the minds of some as to whether or not the vote just taken was legitimate." Everyone was quiet. "For that reason, I submit my resignation so that another vote can be taken." Ruby Doris, who had not been privy to the caucus, did the same. That left John Lewis. And John, who is nobody's fool, immediately realized what was happening. He was angry as hell, too angry.

Shaking with fury, he announced that he would not resign. He claimed that he was chairman, that he had been legitimately elected and that he was not going to allow a group of troublemaking Northerners to take the office from him. These comments generated much more resentment toward John than already existed. They made it appear as if he thought the chairmanship *belonged* to him. They also reminded people of the role he had played in the Selma to Montgomery March and several other instances where he had opposed the majority of the organization's members.

The meeting quickly degenerated into arguing, fussing and cussin'. As each hour passed, as the charges and countercharges grew more virulent, John's support dwindled. By five the next morning, it was obvious that he could do but one thing: resign. He did. In the second vote, Ruby, Stokely and I were elected by a landslide.

Although Stokely was obviously committed to moving SNCC

in a more radical direction, he did not have anywhere near as much power as outsiders believed. The Central Committee, which was composed of veteran SNCC members elected at large, possessed the real power. Stokely wasn't about to change SNCC in any way not approved beforehand by the Committee.

During the first month after the election, Stanley Wise, Stokely and I traveled across the South visiting SNCC projects. Stokely wanted to get a clear idea of the work people were doing. We were in Little Rock, Arkansas, talking with Project Director Ben Greenich and some of his staff when a lawyer came up and told us that James Meredith had been killed.

"Who did it? How did it happen?"

The lawyer didn't know. He had only heard a news bulletin on the radio. "All I know is that some white fella shot him in the head while he was walking down some Mississippi highway," he replied.

The news of Meredith's death reminded me of the dull, aching pain that seemed always to be lurking in the pit of my stomach. Even though I'd always believed that Meredith's intention to march across Mississippi in order to prove that blacks didn't have to fear white violence any longer was absurd, I was enraged.

We didn't find out until two hours later that Meredith had not actually been murdered. The pellets from the shotgun, which had been fired from about fifty feet, had only knocked him unconscious. Although he lost a great deal of blood, doctors in the Memphis hospital where he had been taken were predicting that he would recover. Because we were only a few hours' drive from Memphis, we decided to go there the next day.

When we arrived at the hospital the next afternoon, Dr. King and CORE's new national director, Floyd McKissick, were visiting Meredith. Stanley, Stokely and I joined them. After saying hello to Meredith and congratulating him on his "good luck," we left with Dr. King and McKissick. Meredith was still very weak. On the way down, we were informed that although initially reluctant, Meredith had agreed that the march should be continued without him. He intended to join it as soon as he recuperated.

The decision to continue the march presented us with a dilemma. After the Selma-to-Montgomery fiasco, SNCC had unofficially decided not to be drawn into any more marches—no matter what the circumstances. It was obvious to us, however, that this policy was going to have to be altered. The route that Dr. King and McKissick were mapping out was located in the heart of SNCC territory. As a matter of fact, SNCC possessed more influence in that section of Mississippi than anywhere else in the nation.

Later that afternoon, a group, which included Stokely, Stanley Wise, Dr. King, McKissick and me, drove out on the highway to the spot where Meredith had been ambushed, and we walked for about three hours. We wanted to advertise that the march would be continued. Although we were all quite tense, there was only one incident. A burly white state trooper, who insisted that we could not walk on the roadway, pushed Stokely. In the ensuing scramble, I was knocked to the ground and nearly trampled by Dr. King, who was attempting to keep Stokely from attacking the trooper.

Toward nightfall, we returned to Memphis. Stokely, Stanley and I rode back with Bob Smith, who'd driven up from the Holly Springs SNCC office. We were all of the same mind. SNCC *had* to participate in the march. We also agreed that we could get a lot out of it if we played our cards right. Our only problem was the Central Committee. We knew that we were going to have to be supersalesmen to get them to allow us to join the march.

"As soon as everything is set, we'll fly back to Atlanta and try to get the go-ahead," Stokely told Bob. "In the meantime, I want you to start mobilizing your staff."

Back in Memphis, we got rooms at the Lorraine Motel.*

Early the next morning, Stokely and Stanley and I flew back to Atlanta for a meeting with the Central Committee.

After repeatedly guaranteeing the Committee's members that the march wouldn't cost SNCC anything, we got a very, very reluctant okay. By late afternoon, we were back in Memphis. During the flight, I thought several times about the Committee's last words to us: "We don't want to hear anything more about this march. Don't call us for help!"

Late that night, a planning meeting was held at the Centenary Methodist Church, whose pastor was an ex-SNCC member, the Reverend James Lawson. The meeting was attended by representatives from all those groups interested in participating in the march, including Roy Wilkins and Whitney Young, who had flown in earlier in the day.

Participants in the meeting were almost immediately divided by the position taken by Stokely. He argued that the march should deemphasize white participation, that it should be used to highlight the need for independent, black political units, and that the Deacons for Defense, a black group from Louisiana whose members carried guns, be permitted to join the march.

Roy Wilkins and Whitney Young were adamantly opposed to Stokely. They wanted to send out a nationwide call to whites; they insisted that the Deacons be excluded and they demanded that we issue a statement proclaiming our allegiance to nonviolence.

---

* The motel where Dr. King was assassinated while standing on a balcony in April, 1968

Dr. King held the deciding vote. If he had sided with Wilkins and Young, they would have held sway. Fortunately, he didn't. He attempted to serve as a mediator. Although he favored mass white participation and nonviolence, he was committed to the maintenance of a united front. Most of his time was spent attempting to get the rest of us to agree that unity was necessary. It was obvious to me from the beginning that the possibilities of unity were almost nil.

Toward morning, McKissick, who like the rest of us was tired of arguing, gave up on the possibility of a united front. After announcing his support of whatever position Stokely took, he went to bed. Despite considerable pressure, Dr. King refused to repudiate Stokely. Wilkins and Young were furious. Realizing that they could not change Stokely's mind, they packed their briefcases and announced that they didn't intend to have anything to do with the march. By the time we held the press conference the next day to announce officially that the march would occur, they were on their way back to New York City.

The march began in a small way. We had few people, maybe a hundred and fifty. That was okay. We were headed for SNCC territory and we were calling the shots. We had conducted an all-SNCC meeting with Bob Smith and his staff before the march began and everything was perfectly organized.

A small crew of SNCC organizers had been assigned the task of traveling ahead to make contact in the communities through which we passed. Old SNCC volunteers, people who had worked with us during the summer of 1964, were contacted in each town. They were asked to provide meals, sanitary facilities and sites for the nightly mass meetings. They were also told to make preparations for the voter-registration drives that we intended to conduct in each town.

Those participating in the march were repeatedly told that it was their march and they should feel free to use "any means necessary" to keep it from being disrupted.

"Does that include violence?" we were asked on several occasions.

"Any means necessary is self-explanatory," we would reply.

Although SNCC people were dominating the march, Dr. King was enjoying himself immensely. Each day he was out there marching with the rest of us. His nights were spent in the huge circuslike sleeping tent. For one of the first times in his career as a civil rights leader, he was shoulder to shoulder with the troops. Most of his assistants, who generally stationed themselves between him and his admirers, were attending a SCLC staff meeting in Atlanta.

During the long, hot days as we trekked along the side of the road, we had an opportunity to engage him in friendly conversation. He turned out to be easygoing, with a delightful sense of humor. We had an opportunity during the discussions to present our ideas and our approach to the struggle. His mind was open and we were surprised to find that he was much less conservative than we initially believed.

Little by little, he began to agree that it *might* be necessary to emphasize Black Consciousness. He also agreed that our commitment to independent, black organizations *might* just work. By the end of the first week, he was giving speeches at the nightly rallies in favor of blacks' seeking power in those areas where they were in the majority.

From the very beginning of the march, poor blacks along the route were awestruck by Dr. King's presence. They had heard about him, seen him on television, but had never expected to see him in person. As we trekked deeper into the Delta, the people grew less reserved.

The same incredible scene would occur several times each day. The blacks along the way would line the side of the road, waiting in the broiling sun to see him. As we moved closer, they would edge out onto the pavement, peering under the brims of their starched bonnets and tattered straw hats. As we drew abreast someone would say, "There he is! Martin Luther King!" This would precipitate a rush of two, sometimes as many as three thousand people. We would have to join arms and form a cordon in order to keep him from being crushed.

I watched Dr. King closely on several such occasions. The expression on his face was always the same, a combination of bewilderment, surprise and gratitude. He would smile a little, nod his head in a thank-you gesture and touch as many of the reaching hands as possible. Sometimes we would halt the line of marchers while he delivered a speech, promising that things were going to get better and urging them to register and vote.

It's difficult to explain exactly what he meant to them. He was a symbol of all their hopes for a better life. By being there and showing that he really cared, he was helping to destroy barriers of fear and insecurity that had been hundreds of years in the making. They trusted him. Most important, he made it possible for them to believe that they *could* overcome.

The nightly rallies, mostly conducted in rustic little churches, were beautiful. Long before they began, all seats would be filled. The people wanted to see and hear Dr. King. With shouts of "Amen!" they would urge him to, "Do tell." Their responses inspired him. Each of his speeches was extemporaneous, from the heart. He was getting on with the people and enjoying every minute.

As we got closer to the Greenwood area, the nightly meetings took on the character of a speaking fete. Stokely, who had worked out of Greenwood during the summer of 1964, was well known. Many of those who attended the nightly rallies wanted to see and hear him. Others were attracted by Dr. King. The two of them were like dynamite. Their fervent speeches left all who heard them both emotionally exhausted and inspired.

McKissick, who was generally the first speaker, received a less enthusiastic response. He was better known farther South in the Yazoo City area. CORE people had been working there for years. He was not upset by the enthusiastic response given to Dr. King and Stokely. He understood that people at Yazoo knew his work and loved him; his time would come.

The Deacons for Defense served as our bodyguards. Their job was to keep our people alive. We let them decide the best way to accomplish this. Whenever suspicious whites were observed loitering near the march route, the Deacons would stop them and demand that they state their business. In those areas where there were hills adjacent to the road, they walked the ridges of the hills. We did not permit the news media's criticism of the Deacons' guns to upset us. Everyone realized that without them, our lives would have been much less secure.

We had our first major trouble with the police on June 17, in Greenwood. It began when a contingent of state troopers arbitrarily decided that we could not put up our sleeping tent on the grounds of a black high school. When Stokely attempted to put the tent up anyway, he was arrested. Within minutes, word of his arrest had spread all over town. The rally that night, which was held in a city park, attracted almost three thousand people—five times the usual number.

Stokely, who'd been released from jail just minutes before the rally began, was the last speaker. He was preceded by McKissick, Dr. King and Willie Ricks. Like the rest of us, they were angry about Stokely's unnecessary arrest. Their speeches were particularly militant. When Stokely moved forward to speak, the crowd greeted him with a huge roar. He acknowledged his reception with a raised arm and clenched fist.

Realizing that he was in his element, with his people, Stokely let it all hang out. "This is the twenty-seventh time I have been arrested—and I ain't going to jail no more!" The crowd exploded into cheers and clapping.

"The only way we gonna stop them white men from whuppin' us is to take over. We been saying freedom for six years and we ain't got nothin'. What we gonna start saying now is Black Power!"

The crowd was right with him. They picked up his thoughts immediately.

"BLACK POWER!" they roared in unison.

Willie Ricks, who is as good at orchestrating the emotions of a crowd as anyone I have ever seen, sprang into action. Jumping to the platform with Stokely, he yelled to the crowd, "What do you want?"

"BLACK POWER!"

"What do you want?"

"BLACK POWER!!"

"What do you want!?"

"BLACK POWER!! BLACK POWER!!! BLACK POWER!!!!"

Everything that happened afterward was a response to that moment. More than anything, it ensured that the Meredith March Against Fear would go down in history as one of the major turning points in the black liberation struggle. The nation's news media, who latched on to the slogan and embellished it with warnings of an imminent racial cataclysm, smugly waited for the predictably chaotic response.

Dr. King's assistants were among the first to react. They rushed in and demanded that he dissociate himself from Black Power, no matter what its meaning. They insisted on a "change in emphasis" for the march. They demanded that more whites be brought in and that we stop using Black Power for the duration of the march.

Although he disavowed Black Power, Dr. King did not follow all the advice of his assistants. They wanted him to repudiate Stokely and SNCC, but he refused. When they realized that Dr. King was not going to take their advice, his associates took another tack. They began to talk about the need for better organization. Hosea Williams, one of SCLC's organizers, was put in charge of advance accommodations. This had been Bob Smith's job. I was sent to Jackson to work with the committee that was to organize a gala rally for the last night of the march. Other SNCC people had their jobs taken from them, too.

Despite the opposition of Dr. King's assistants, Charles Evers, James Meredith and the nation's news media, we managed to keep Black Power alive. We did not intend to be maneuvered out of the picture as we had been in Selma and at the March on Washington. Although outnumbered, we managed to remain one step ahead of our critics at every point.

Stokely was largely responsible. He used every opportunity to shout the need for Black Power. At each interview, rally and press conference, he kept hammering away. The crowds that came out to meet us in each town were being primed by Bob Smith's staff. Whenever Stokely mounted a platform and asked "What do you want?" they responded without hesitation: "BLACK POWER! BLACK POWER! BLACK POWER! BLACK POWER!" During such moments, Bob Smith, Willie Ricks and I moved through the crowd distributing Black Power leaflets and placards. We had had them printed the day after the Greenwood rally at the SNCC print shop in Atlanta.

We had several run-ins with the police after Greenwood, but we didn't allow them to alter the thrust of the march. On at least two occasions, they attacked us with tear gas and clubs. This didn't destroy our enthusiasm. The crowds got larger and larger. They all wanted to hear about the same thing: Black Power.

James Meredith attempted at one point to take over the march by striking out ahead of us. This didn't work. We told him that he could go off on his own if he wanted. It was clear, however, that if he did, he would be by himself. The people were with us. Meredith had been outgrown by his own march.

From SNCC's point of view, the march was a huge success. Despite the bitter controversy precipitated by Stokely's introduction of Black Power, we enjoyed several important accomplishments: thousands of voters were registered along the route; Stokely emerged as a national leader; the Mississippi movement acquired new inspiration, and major interest was generated in independent, black political organizations.

One of our most important accomplishments was the deep friendship that developed between Dr. King and those SNCC members who participated in the march. I have nothing but fond memories of the long, hot hours we spent trudging along the highway, discussing strategy, tactics and our dreams.

I will never forget his magnificent speeches at the nightly rallies. Nor the humble smile that spread across his face when throngs of admirers rushed forward to touch him. Though he was forced by political circumstance to disavow Black Power for himself and for his organization, there has never been any question in my mind since our March Against Fear that Dr. King was a staunch ally and a true brother.

CHAPTER 13

A Political Nightmare: A Personal Tragedy

IT IS DIFFICULT, virtually impossible, for me to recount the garishness of the nightmare that dominated our lives after the Meredith march. The hysteria-ridden responses that Black Power generated are still too much for me to comprehend. It seemed impossible to pick up a newspaper or magazine or listen to the radio without encountering a warning about the dangers of Black Power. Two important attacks were made at the NAACP's national convention in early July. At the convention, which was held in Los Angeles, Roy Wilkins and Vice-President Humphrey blasted us.

"No matter how endlessly they try to explain it, the term 'black power' means anti-white power," Wilkins charged. "It has to mean 'going it alone.' It has to mean separatism. . . . We of the NAACP will have none of this. We have fought it too long. It is the ranging of race against race on the irrelevant basis of skin color."

Vice-President Humphrey, our old nemesis from Atlantic City, was one of the featured speakers. Because of his office and his alleged commitment to civil rights, his attack proved much more damaging than that of Wilkins. His speech, which has been called a "masterpiece of identification of the speaker and what he stands for with the audience and what it stands for," was, at best, conscious manipulation and, at worst, stupid:

> It seems to me fundamental that we cannot embrace the dogma of the oppressor—the notion that somehow a person's skin color determines his worthiness or unworthiness.
>
> Yes, racism is racism—and there is no room in America for racism of any color.
>
> And we must reject calls for racism, whether they come from a throat that is white or one that is black. . . .
>
> We must strive to create a society in which the aims of the National Association for the Advancement of Colored People and the civil rights movement can be achieved. And always remember, we seek *advancement* . . . not apartheid. . . .
>
> Integration must be recognized as an essential *means* to the *ends* we are seeking.

I remember thinking two things as I listened to the radio report of Humphrey's speech: that it was at least two years too late and that it was addressed to the wrong group. He should have delivered it in Atlantic City at the Democratic National Convention. He should have said there the same things to his colleagues, to the admittedly racist Mississippi delegates whom he endorsed while rejecting those from the MFDP.

No matter how many speeches we gave in order to set the

record straight, we were unable to convince most Americans that we weren't interested in sacking cities or dragging white women off to the Black Belt to be gang-raped by black fiends. The number of critics echoing the unfounded charges of Roy Wilkins and Hubert Humphrey grew more numerous with each passing week.

The critical condescension of New York *Post* columnist James Wechsler, who called Black Power "a diversionary disaster," was typical.

"There is a large, ghastly national tragedy unfolding," Wechsler charged in the December issue of *The Progressive.* "The 'Killers of the dream' are on the offensive, and some are those who not only shared the dream but have endured great personal risk in the struggle . . . the evidence is unescapable that the cause of civil rights is floundering, all the visions of the Freedom Movement are imperiled, and some deeply dedicated men are helping to set the stage for the destruction of the noblest cause of our time."

I was tempted after reading Wechsler's article to write him a letter saying, *"I'd rather be FREE and IGNOBLE than oppressed and noble."*

One of the few bright spots during this period was a wholly unexpected statement of support from the National Committee of Negro Churchmen. The statement, which took up a full-page spread in the July 31 issue of *The New York Times,* was an eloquent surprise.

"The fundamental distortion facing us in the controversy about 'black power' is rooted in a gross imbalance of power and conscience between Negroes and white Americans," the Churchmen charged. "It is this distortion, mainly, which is responsible for the widespread, though often inarticulate, assumption that white people are justified in getting what they want through the use of power, but that Negro Americans must, either by nature or by circumstances, make their appeal through conscience."

While most critics were claiming that SNCC was intent on replacing white demagogues with black ones, the Negro Churchmen spoke forthrightly to the real issues. "As black men who were long ago forced out of the white church to create and wield 'black power,' we fail to understand the emotional quality of the outcry of some . . . against the use of the term today. It is not enough to answer that 'integration' is the solution. For the issue is not one of racial balance but of honest interracial interaction.

"For this kind of interaction to take place, all people need power, whether black or white."

No one has better capsulized the need for black power. Unfortunately, few of those who addressed themselves to Black Power were as honest or as forthright. For the most part, we saw only statements of criticism. By the year's end, it was obvious that a frightening percentage of the nation's citizens were convinced that SNCC's members were irrational brutes bent on senseless mayhem.

The criticism and the pressure that accompanied it had profound effects on our personal lives. As the days and weeks passed, I felt more and more like a yo-yo, spinning crazily on the end of a taut string. If this hadn't been the case, things might have turned out differently between Sandy and me.

I met Sandy while in Jackson, Mississippi, organizing the big rally for the last night of the Meredith march. She was tall and slender with a bright engaging smile. Her skin was honey-brown and she moved with the grace of a cheetah. I was immediately attracted to her. Most important, our personalities seemed compatible. I felt no need to play Mr. Masculinity games and she didn't resort to Miss Hard-to-Get tactics.

When my work in Jackson was done and it was time for me to return to Atlanta, I asked her to come with me. "I can't offer you much. I don't have much, only myself and what I feel for you. If you come, I think we can make a life together. I really want you to come with me."

Although she was only eighteen, she was mature enough to understand the feelings we shared and the meaning of my request. "Give me a couple of weeks to clear up a few things," she replied. "As soon as I've taken care of them, I'll come."

By the end of July, we were living together in my crowded little apartment in southwest Atlanta. We were ecstatically happy the first few weeks. Then, reality began to intrude. Our problem was simple: I couldn't steal enough time from SNCC to establish the kind of rapport with her that a man and woman need in order to live together successfully.

Like a demanding mistress, SNCC drained vital energy from me. There was always a crisis of some sort, some meeting to attend, some urgent problem to be solved. We never had time to indulge our relationship. SNCC left us no privacy. It was always there, always lurking in the back of my mind. By the end of our first two months together, I was struggling with a terrible personal dilemma: every moment with Sandy was a moment I spent away from SNCC.

"Maybe it will be easier if I join SNCC," she suggested one night in a fit of exasperation. "If I went to the office with you each day, at least we could see each other. We might be able to arrange lunch breaks at the same time." There was no way for me to explain that I didn't take lunch breaks.

"That won't work," I replied. "The organization would still come between us, probably more so. If you did that, we would probably end up with no time to ourselves. We'd probably spend all our spare time discussing SNCC business. Let's try to wait it out. This uproar over Black Power can't last much longer. Then I'll be able to spend more time with you. Then we'll be able to have a personal life together."

I was wrong. Things did not cool down. It was the most amazing thing I had ever seen. Each week, the controversy over Black Power grew larger. We were mercilessly hounded by newsmen, some of them coming from as far away as Japan. It seemed at times to be an inexorable pull radiating out from Black Power. Every SNCC program and every SNCC person was drawn toward the term's controversial vortex.

By September, it looked as if we were finally going to get a respite. Even the weather was cooling down. The controversy seemed to be ebbing. Wrong again. On Tuesday, September 6, a white police officer in Atlanta shot and wounded a black man who was suspected of having stolen a car. The suspect, whose name was Harold Prather, was considerably slowed by the first shot. The officer could have caught and subdued him with little effort. Instead, he leveled his gun and shot Prather a second time. When he was finally arrested, Harold Prather was lying unconscious in a pool of blood on the porch of his mother's home.

Within minutes after the shooting, a small, angry crowd of local residents gathered in the street in front of Prather's mother's home. Most of the crowd's members had seen the shooting; others had heard about it through the grapevine. They were all angry.

While the crowd was milling about in the street trying to decide what to do, Stokely came on the scene. He was with Emmanuel Hall, a local news commentator for a black radio station. Hall had just finished conducting an interview with Stokely and was driving him back to the SNCC office.

After being informed of the shooting, Stokely told the angry people that he would return later. By the time Stokely arrived at the office, we had heard about the shooting on the radio. When he announced his intention of returning to the scene, we cautioned him against doing so.

He insisted, however, that SNCC should send someone in his place. We decided to send Bill Ware and Bobby Walton because they were both experienced and we knew that they could take care of themselves. Moreover, the police did not know either of them. We only had time to give them minimal instructions: "Take the sound car. When you get there, size up the situation and say what you think is appropriate."

By the time Bill and Bobby arrived at the scene of the shooting, the crowd was larger. Like most of the people, Bill and Bobby were convinced that Harold Prather was dead.

"Let's stop traffic. Let's do something about them killing black people," Bill shouted through his microphone. He didn't get to say much more because he was promptly arrested for operating a sound system without a permit and inciting to riot.

During the next five hours, the people in the surrounding community threw bricks at police officers, tried to free an arrested man and attempted to overturn a police wagon. No stores were looted and no one was seriously injured.

At one point, Atlanta's mayor, Ivan Allen, appeared on the scene. When he asked the people what they wanted, they demanded that he remove the police from their neighborhood. While Mayor Allen was talking to the crowd, a group of angry teen-agers shook the car on which he was standing. Although he lost his balance and almost fell, the mayor wasn't injured.

Despite the mildness of the disturbance, Atlanta city officials and the local news media acted as if the city had been subjected to a major Watts-style rebellion. The attitude expressed by Eugene

Patterson, then editor of the Atlanta *Constitution*, was representative. Patterson's description of Mayor Allen in action read like the script of a third-rate soap opera.

"He lifted his reddened eyes to the porches and looked at the Negro men and women and children whose rights he had long fought for at the risk of his political life and sighed, 'They don't know, they just don't know.'"

Patterson did not print *everything* Mayor Allen said, however. For after his paternalistic attempts to get the people to return to their homes proved unsuccessful, the mayor turned to nearby police officers and said, "Get them out of here, if you have to tear it down brick by brick!"

The mayor, who had a reputation for being such a "good liberal," was not referring to the entire city. Rather, he suggested that police might need to destroy the mean little shanties that housed Harold Prather's justifiably angry neighbors.

City officials and the news media played down the fact that the disturbance was initiated by a white police officer's shooting an unarmed black man. They attempted instead to create the impression that SNCC was responsible for the entire affair.

"Like the old white mobs, the rock-throwing Negroes didn't have any clear idea what had hold of them Tuesday," wrote Eugene Patterson. "Demagogues had hold of them; SNCC was in charge."

Stokely was arrested at SNCC headquarters the next afternoon by a tight-jawed contingent of Atlanta police officers. He was charged with inciting to riot and with disturbing the peace. His bond was set at ten thousand dollars.

Few would listen when we attempted to explain that Stokely wasn't even on the scene when the trouble occurred. He and SNCC were considered to be the source of some dangerous plague. We were getting it from all sides. One hundred students from an all-black high school signed an anti-SNCC petition and gave it to the press. A black minister called us an "albatross" around the neck of the black community. All we could do was continue to state the truth and hope that the storm would soon subside.

The storm didn't subside anywhere near as soon as we had hoped. While a group of influential Georgians were pressuring the city of Atlanta to charge Stokely with insurrection, a capital offense in Georgia, Congress began to get into the act. Representative Wayne Hays from Ohio took to the floor of the House and called for a federal crackdown on Stokely, whom he accused of "inciting to riot in Atlanta and Cleveland, Ohio."

"Carmichael and his anarchist group belong behind bars and the quicker we get him there, the better off this country is going to be," Hays charged. Newspaper reports indicated that his comments were greeted by a "round of applause."

Stokely was so upset by the untrue charges being lodged against him that after he was free on bail he spent the entire day knocking on doors in the neighborhood where the disturbance had occurred. He explained to the people just what he did on the day of the trouble.

The Atlanta disturbance and the responses it generated added greatly to my work at the office. I was turning in sixteen-hour workdays as a matter of course. I knew that Sandy needed me, but most of the time I was too tired when I arrived home to give her anything more than a perfunctory hug and kiss. Many nights, I fell asleep in the middle of a conversation, leaving her awake and alone. I knew that trouble was brewing, but I didn't know how to avoid it.

Late in September, I had to make a two-day trip for SNCC. When I returned home, Sandy wasn't there. After waiting an hour for her to return, I got worried and went looking for her. I found her sitting in a nearby restaurant having a drink with Rap Brown. They didn't see me and I didn't disturb them. When she got home a few hours later, I asked her where she'd been.

"I was shopping with girlfriends," she said.

I didn't say anything. My hope was that she could come to understand what she was doing and tell me the truth. A few weeks later, I had to go on another trip for SNCC. One of my stops was the SNCC project headquarters in Lowndes County. As soon as I stepped out of my car, one of the staff people came up and told me he had just seen Sandy!

"Where?"

"She just went through here with Rap. They were apparently on their way over to Greene County in Alabama. I think that Rap is working over there."

When I got back to Atlanta a few days later, Sandy was back home. I gave her every opportunity to mention her trip with Rap, but she didn't. After a couple of days, I couldn't take it anymore. I confronted her with my knowledge. She didn't lie. Although she was embarrassed, she admitted that she and Rap were having an affair.

"Cleve, I need more than I'm getting from you. The organization is taking everything. I mean, I'm scared. I never know when you leave here if you'll return. You could be killed any minute. I can't do anything about it. I love you. You know that. At the same time, I am being consumed by my own fears!"

"Well, goddamn, Sandy!" I yelled. "Fuckin' Rap isn't going to solve anything. *He* knows that. You are not the only one who's afraid! Shit. Do you think I relish this situation? It fucks with me just as much as it fucks with you!"

A painful expression spread slowly across her face. Tears welled in her eyes. She began to twist the strap of her purse with nervous fingers. She was touching it absently. It was an object that happened to be near enough to grasp, to hold on to. She seemed so vulnerable.

"I'm sorry," I continued, my voice now calm. "I didn't mean to yell at you. I understand how you feel. I know it's hard. You aren't the only one who has drifted into such a relationship. Hell, it happens all the time, every day." This calmed her a bit and she stopped twisting the purse strap.

"Please don't cry. Just try to hang on a little while longer. Things will get better. I'm going to take some time off, maybe a month or two. We'll get away together, leave Atlanta. We'll do all the things we've been planning."

I put my arm around her and as I tried to console her, I could feel her shoulders shaking. She made no sound, but large tears spilled down her cheeks, wetting her face and her dress.

"It's gonna be all right, honey," I promised. "You'll see."

And I did manage to curtail my work at the office during the next three or four days. But by the end of the week I was back in the same old rut. No matter how well I planned my time, there seemed always something to keep me at the office late into the night. I just couldn't seem to break the gruesome work-sleep, work-sleep cycle that had dominated my life for so long.

In late December, SNCC held a staff meeting at a resort in upstate New York. We had a lot of business to take care of and I was kept busy most of my waking hours. I was only able to be with Sandy during mealtimes. Because I was tense and tired, I didn't even feel like talking when I saw her during these meals.

Then one afternoon, I was sitting in one of the interminably long workshop sessions and it occurred to me that I had not seen Sandy all day. It also occurred to me that I had not seen Rap, either. Leaning over, I whispered to Bob Smith, who was sitting next to me.

"Bob, I think Sandy's with Rap again."

"Man, what do you want *me* to do about it?" he replied.

"I want you to find them and tell her to meet me in our room."

I asked Bob to help me because I didn't want to precipitate a crisis in SNCC. It sounds strange now that our personal lives were so intimately tied to SNCC. SNCC was my life. I had to be as discreet as possible. Sending Bob for Sandy was the only thing I knew to do. A terrible scene would have occurred if I had confronted Sandy and Rap together. A confrontation with Rap would have led inevitably to a split between those who supported me and those who supported him. I didn't want that to happen.

When Sandy got to the room, I sat beside her on the bed and tried to explain my thoughts about her affair with Rap. I told her that her behavior could help or destroy the unity in the organization. I realized almost immediately, however, that most of what I was saying probably didn't make any sense to her. She wasn't a SNCC person. She hadn't gone through what I'd been through. Names like Herbert Lee, Wayne Yancey and Sammy Younge didn't mean to her what they meant to me.

"Sandy, I know this probably won't make sense to you, but you've got to stop this. You and Rap *are* threatening the entire organization. If you continue, there's going to be trouble, big trouble."

Her eyes told me that she didn't really understand how her actions threatened SNCC, but she did agree to discontinue her affair with Rap. I was relieved. But what I did not realize at that time was that we—all of us connected with SNCC—were responding to forces beyond our control.

On the way back to Atlanta, we stopped off in New York City

to attend a New Year's party hosted by Stokely's mother and sisters. The party was held at Stokely's home in the Bronx. Because the meeting at the resort had been productive, there was an air of conviviality at the party. Everyone was dancing, laughing and having a good time.

At one point during the evening, I left my seat beside Sandy and went to the bathroom. When I got back, Rap was sitting in my seat. He and Sandy had their heads together, talking quietly.

I stood in the doorway watching them, trying to control the rage that swelled inside me. After nearly five minutes, Sandy looked up and I caught her attention. I beckoned to her and turned to go outside into the cool night air. I was furious.

I didn't give her a chance to say anything, to explain, to justify. As soon as she was close enough, I lashed out, slapping her hard across her face. And I stood there in the dark, beating her . . . in the face, in the chest, on the back of her head, wherever my hands landed. Twice, the force of my blows nearly knocked her from her feet. By the time I was able to stop, her face was puffy, swollen. Her hair was wildly tangled. She was crying uncontrollably. I spoke to her through clenched teeth.

"I don't want any more of that shit, Sandy!" My voice sounded hoarse, strange, as if someone else were speaking. "Goddammit, do you understand? Don't do this again! Don't fuck over me again! Do you hear what I say?"

She looked straight at me, her tears falling and filling her eyes again. She nodded, but didn't speak.

"All right! Now, we're going back inside. We're going to stay a few minutes and then we're going to leave. Now let's go!"

I realize now that I was terribly unfair to Sandy at the time. But, then I thought she was being unfair to me. I thought she was being unfair to SNCC. I rationalize that I struck her because I was angry, because I was hurt, because I was frustrated, because I was tired. Very, very tired. I honestly didn't know what else to do.

I never said a word to Rap about Sandy. This doesn't mean I was not bothered by his actions. More than anything else, I pitied Rap. I thought him weak. It didn't take any great ability on his part to win the confidence of someone who was in such need of solace.

Sandy left me in late February. We both agreed that it was probably the best thing for her to do. There were no angry words or bitter denunciations. We agreed to keep in touch and maybe we would even make another go at it sometime in the future.

What happened between Sandy, Rap and me was not unique. SNCC people were continually struggling with similar problems. During the early portion of the organization's history, we managed to keep our problems under control. But with the conclusion of the Meredith march, nearly everyone experienced some personal tragedy brought on by commitment to the organization and the principles for which it stood.

CHAPTER 14

"The best of times: the worst of times . . ."

BLACK POWER THRUST SNCC to the forefront of the struggle for black liberation. Although SCLC, CORE, the NAACP and the Urban League continued to have prestige, SNCC was the premier organization. Journalists, intellectuals, politicians, students: they were all responding to our radical cadence. In black ghettos across the nation, we were hailed as heroes—Young Turks taking it to "The Man."

The adulation and constant attention were only part of it. For during this period, SNCC was wracked by massive internal problems. Dissension, hostility, confusion and personal tragedy dominated the private SNCC.

We were surrounded by contradictions. While Establishment leaders denounced our spiraling influence, we were doing everything possible to keep the organization from collapsing. While Stokely, Jim Forman and several other SNCC members crisscrossed the nation, making Black Power speeches to enthusiastic crowds, the organization's membership was steadily declining.

Although a sizable portion of the nation's citizens were convinced that SNCC was an elaborate political juggernaut with representatives and branch offices strewn across the nation, such was not the case. Our support of Black Power and criticism of the Vietnam War virtually eliminated all our funding sources. By December, 1967, SNCC had fewer than ten offices in full operation. And one of our most difficult tasks was keeping the seventy-five to eighty people still on staff fed, housed and clothed.

Despite our problems, we were very successful in getting our ideas to the people we most wanted to reach—poor blacks. This would not have been possible if it hadn't been for the news media. Stokely was considered "good copy" and the media were eager to record his every word. It didn't take long for us to realize that by calling a press conference and making a statement, Stokely could reach millions of people.

Although the newsmen who interviewed him frequently were hostile, Stokely—whose mind is extremely agile—generally managed to manipulate them to his advantage. Unlike some speakers, who have to warm up to a topic by talking for several minutes, Stokely can move to the heart of a problem and state his position immediately.

Stokely's magnetic appeal to the mass media presented SNCC with several problems. Unlike the organization's previous chairman, Stokely was outgoing, a charismatic figure who attracted attention wherever he went. Whenever he appeared on the street, crowds would form. Complete strangers, black and white, would ask for his autograph. For the first time in SNCC's history, its members were regarded by the public as followers.

Several members resented this. After repeatedly seeing themselves portrayed on television and in the newspapers as Stokely's subordinates, they began to resent him. The term "Stokely Starmichael," which had been used to kid him during the Mississippi summer, was reactivated. Though I spent a great deal of time and energy talking with members of the organization, trying to get them to place Stokely's notoriety into a proper perspective, I was not completely successful.

During the first months of 1967, a serious rift developed in SNCC over the operation of the Atlanta Project. The Project, which was located in a gruesome slum called Vine City, was initiated shortly after Julian Bond's election to the Georgia House of Representatives. Because it was located in Julian's district, we had hoped to help him organize a community-based political movement. We intended to use the Project to prove to skeptics that Black Power concepts could solve the problems of poor blacks in urban slums.

The Project was headed by Bill Ware, a veteran SNCC member, and several organizers previously active in Mississippi and Alabama. Shortly after Stokely was elected, the problems began. Bill and his staff refused to take orders. They ignored memos, refused to return phone calls and rarely attended general staff meetings.

By January, 1967, the Atlanta Project staff was on the verge of demanding complete autonomy. Although this Project was completely supported by funds coming from the National Office, the staff adamantly refused to work with the organization's leadership. Things came to a head when they refused to return a Sojourner Motor Fleet car that had been reassigned to the printing office. When a series of polite letters from Stokely and me were ignored, I was instructed by the Central Committee to take stronger actions.

On February 3, I wrote Bill Ware a letter informing him that the entire Atlanta Project staff had been temporarily suspended until the Central Committee had an opportunity to review the Project's operation. I knew that trouble was in the offing when Bill and his staff ignored the suspension by continuing to work as if they hadn't received the letter.

When Stokely and I went to the Project's headquarters, which was located in a three-room "shotgun" shanty, we received a hostile response. No one would even tell us where the car was. Bill Ware, who did most of the talking, was obviously jealous of Stokely. Although he sensed this, Stokely was reluctant to force the issue.

"I think we ought to take it easy, give them an opportunity to calm down, then we can talk to them," he told me during the short ride back to the National Office, which was located less than a mile from Vine City.

"Look, Stokely," I replied, "if those guys go out and wreck that car, SNCC could end up owing thousands of dollars. I know that you don't want to push them into a corner, but there are larger issues involved. I think we ought to report the car as missing. I don't see anything else to do. We can't even afford to pay the few people still on staff enough to feed themselves. This is business. If Bill and his staff want to play games, let them. We have a responsibility to SNCC to move on this issue. They've been getting by long enough."

Later that day, we went down to police headquarters and reported the car missing. The Atlanta Project staff responded by firing off a threatening telegram to Jim Forman, who was in New York trying to get his ulcers under control. The telegram, indicative of the distrust and bitterness that were beginning to destroy SNCC from within, was signed by Bill Ware:

> Your hand-picked Chairman, the alleged hope of Black America in the calculated conspiracy to destroy the black ideology symbolized by the Atlanta Project, has descended to the level of calling a racist henchman cop of the white master Allen of Atlanta to settle an internal dispute between the supposedly black people of SNCC.
>
> Beware of going to the man to deal with supposedly internal conflicts. It can work both ways. We have tapes and other information that could fall into black people's hands across the country.
>
> There are several magazines lined up to publish our writing.

Stokely found out about the telegram when Jim called the National Office in order to learn what was going on. He responded by writing Bill and his staff a one-sentence letter: "You have been fired from the Student Nonviolent Coordinating Committee."

I had my secretary, Fay Bellamy, prepare a special report for the staff containing copies of every letter sent to the Project staff and a long statement on the reasons for the firing. After reading the report, most of the staff agreed that the situation was handled in a reasonable manner.

The suspension of the Atlanta Project was a crucial loss. SNCC was not in a position to restaff the Project. Everyone on staff was working to capacity and we just didn't have enough money to hire new people. This resulted in SNCC's not having *any* urban project when it was most necessary that we demonstrate the utility of Black Power in an urban setting.

At the time it was generally assumed, and I believe rightly so, that Black Power would work in rural areas where blacks were in the majority. Although we'd lost the election in Lowndes County, we knew that the county's blacks *could* take over the county government. It was just a matter of getting properly organized. It wouldn't be easy, but it was clearly possible.

We also believed that Black Power concepts could be used to organize ghettos, especially the ghettos of the North. As we saw the situation, the most serious obstacles to successful Black Power offensives in urban areas were the hidden sources of power. Power is readily apparent in rural areas: the sheriff, the mayor, the tax assessor, the judge, the few rich families that control decision-making. And when those in power act in a discriminatory manner, everyone knows it.

At the same time that we were desperately attempting to find a way to establish at least one Black Power project in an urban ghetto, SNCC's members were becoming increasingly aware of the international implications of domestic black oppression. Malcolm X had a lot to do with this new awareness. Although we didn't have much personal contact with him, his ideas about the international struggle for human rights made a big impression on our thinking.

Throughout the latter portion of 1966, Malcolm's speeches were frequently discussed at SNCC gatherings. Before his assassination most of us were convinced that his awesome charisma and brilliant insights would have resulted in his becoming one of the first men in history to lead a multicontinental revolutionary movement.

In addition to our discussions of Malcolm's ideas, we spent a lot of time studying the tactics and theories of revolutionaries in Asia, Africa and Latin America. We were particularly attracted to Ghana's Kwame Nkrumah and Algeria's Frantz Fanon. Their ideas about violence and neocolonialism helped us understand many things that had been puzzling us since Atlantic City.

By the beginning of 1967, the majority of SNCC's members considered themselves part of an emerging Third World coalition of revolutionaries who were anticapitalist, anti-imperialist and antiracist. Despite our commitment to internationalizing the struggle, most of SNCC's members believed that we should remain primarily concerned with domestic problems.

We realized that we had to do two things in order to accomplish our objectives: develop new support groups and establish new funding sources. Despite our prominence, we weren't receiving any organized support. We sometimes went as long as five weeks without being able to pay the staff. Many members had to take part-time jobs in order to pay their bills. For those members who couldn't find outside jobs, we had to take up collections in the office. With the money collected, we would buy wieners and beans. These were cooked in a huge pot on a stove in the rear of the office.

After much discussion, we decided that we should turn to students on Negro college campuses in order to win financial support. There are more than one hundred Negro colleges in the nation and we were convinced that we could get substantial support from their students if we approached them properly. As program secretary, I spent a major portion of my time during the first three months of 1967 discussing ways in which SNCC could reestablish contact with Negro college students.

Sometime near the end of March, we created a special position, campus coordinator. If I am not mistaken, George Ware, a SNCC member who had attended Tuskegee Institute and who knew Negro colleges well, was the first person to hold this position. In any event, the coordinator was given a high priority go-ahead and was instructed to organize campus SNCC chapters capable of accomplishing four objectives: (1) raising funds for SNCC; (2) gain-ing political power in their areas; (3) carrying out those programs which interested them and those programs formulated by the parent body and (4) providing "aid and comfort" for SNCC.

Little did we suspect when we initiated the campus-chapters program that it would culminate in a series of bloody confrontations between students and cops. I certainly didn't suspect that one of those confrontations would bring me closer to death than I have ever been.

At this time I was worrying about the collapse of my relationship with Sandy, Stokely's relationship with the rest of SNCC's members and the demise of the Vine City Project. I was also fighting to keep from being called into the army. This fight began in February, 1967, when I found out that the local board in Bamberg, South Carolina, was preparing to draft me. In order to counteract the board, I had Howard Moore, SNCC's attorney, file a petition on my behalf.

The petition, which was designed to block my induction both in Georgia, where I was living at the time, and in South Carolina, was based on four objections: (1) that I had been ordered for induction out of turn because I was a member of SNCC and a civil rights activist; (2) that all draft boards in Georgia and South Carolina were invalid because blacks were systematically excluded from participating in the election process; (3) that medical documents proving the existence of a cardiac defect had been ignored by draft boards in both states and (4) that the actions of the boards deprived me of my constitutional rights—specifically, the right of due process of law and equal protection under the law granted by the Civil Rights Act of 1964 and the Voting Rights Act of 1965.

While my petition was still being processed by the courts, I was ordered to report for induction at the Atlanta Center at 7 A.M. on May 1, 1967. Even though I knew that I would be subject to five years in prison and a thousand dollar fine for what I intended to do, my mind was made up. Under no circumstances did I intend to wear a uniform of the United States military. Stokely, who provided me with moral support while I was making my decision, was with me when I arrived at the Atlanta Center.

"Don't let them get to you," he whispered as I left him at the entrance and walked inside.

It took the induction officials longer than usual to process me because I told them that I knew Khaleel Sayyed. Khaleel, whom I hadn't seen since we were working in Cambridge, Maryland, had been convicted—as a result of the testimony of an undercover police officer—of plotting to blow up the Statue of Liberty.

I was closely questioned by several short-haired officials from the army's counterintelligence corps during most of the morning. By midafternoon, they concluded that I was not a "security risk."

At 3:30 in the afternoon, I was escorted into the "ceremony room." There were five white and five black inductees with me. All of them looked worried. When it was time for us to take the fatal step forward, I was left standing in my tracks. One of the other inductees looked back at me, but the rest stared straight ahead.

After giving me an opportunity to change my mind, the visibly angry induction officers told me that I was free to go. When I got outside, the sun was shining brightly. It felt as if a heavy load had been suddenly lifted from my shoulders. Before I got two steps from the door, Stokely ran up to me and reached out to shake my hand.

"How was it?"

"I'm glad this part of it is over."

Before I could say anything else, we were swamped by a crowd of television and newspaper reporters. During the short interview, Stokely pointed out that in Georgia and South Carolina combined only 6 of the 670 local induction board members were black. At the conclusion of the interview, I issued the following statement:

> The central question for us is not whether we allow ourselves to be drafted, for we have resolved that this shall not happen by any means. But rather the central question for us is how do we stop the exploitation of our brothers' territories and goods by a wealthy, hungry nation such as this. I am committed to give support to my brothers in Vietnam as they fight to keep America from taking her tungsten, tin and rubber. I shall be prepared to support my brothers in Iran when they move to overthrow their puppet regime which gives that country's rich oil deposits to the U.S. I shall be prepared to back my brothers in the Congo when they tell the U.S. "Hell, no! This copper belongs to me." I shall stand ready when my brothers in South Africa move to overthrow that apartheid regime and say to the U.S. "This gold, these diamonds and this uranium is ours." I shall stand with my brothers in Latin America when they throw out American neo-Colonialist forces who would take the natural resources of Latin America for themselves and leave my brothers in utter starvation and poverty.
>
> I shall not serve in this Army or any others that seek by force to use the resources of my black brothers here at the expense of my brothers in Asia, Africa and Latin America. . . .

Less than two weeks after I refused to be drafted, SNCC held its annual election. It resulted in my being replaced by Ralph

Featherstone as program secretary and in Rap's replacing Stokely as chairman.

Rap was Jim Forman's candidate. He probably wouldn't have been elected if Jim hadn't backed him, and his election meant that Jim was back in power. During the year Stokely was in office, Jim had lost a great deal of influence—primarily because of his irresponsible actions in Montgomery. The fact that Stokely was independent and didn't seek his advice very often was another reason for his diminished influence. If Stokely had chosen to run for a second term, he probably would have been reelected.

Stokely's reasons for not seeking a second term were simple: he wanted to return to organizing in order to prove that Black Power would work in the black ghettos of the North. In addition, he was really afraid that he would become further isolated from the rest of the organization if he served a second term.

There was little reason for me to run for a third term with Stokely out of office. The two of us were so closely identified that my effectiveness without him would have been greatly impaired. Both Rap and Jim understood that I was a Stokely supporter.

My personal feelings toward Rap, as a result of the affair he'd had with Sandy, also figured in my decision not to seek a third term: I had no desire to exacerbate the hard feelings between us.

There was at least one other reason why I decided not to run for a third term. Like Stokely, I, too, wanted to get back into the field to begin organizing again. I was tired of attending meetings, giving speeches, traveling about, settling arguments and swinging the organizational hatchet. Unlike Stokely, I wanted to organize in the South, in South Carolina, the place from whence I came.

CHAPTER 15

The End of an Era

SHORTLY AFTER RAP'S election, the Central Committee was called upon to make a crucial decision. Bob Zellner, SNCC's first white field secretary, petitioned the Committee for active status. He had been on leave for two years while working with the Southern Conference Education Fund (SCEF), a basically white organization concerned with problems of poverty and racism in the South.

The petition indicated that Bob wanted to organize a project for poor whites in New Orleans. His wife would be working with him. They had already raised money for the project and intended for it to be entirely self-supporting. Bob made it perfectly clear, however: he wanted to operate the project as a SNCC member.

If Bob had been black with a desire to work with black people, there wouldn't have been a problem. The Central Committee would have okayed his petition and wished him well. Bob knew this. That's why he went through the trouble of submitting a formal petition. He was attempting to force the organization's hand in relation to a staff decision made at the big New York staff meeting in December, 1966.

At that meeting, which was repeatedly disrupted by heated arguments and tears, the staff decided that the five whites remaining in the organization should be fired. It was also decided that those whites who desired to continue working with SNCC after they were fired would be permitted to do so only on a "voluntary contractual basis."

Although these were cold decisions, we felt we had to make them. They constituted an attempt to eliminate the contradictions between SNCC's integrated staff and its demand for unilateral black control in black communities. Although the decisions bothered some members of the organization, they did not develop into a critical issue until Bob submitted his petition to the Central Committee.

Bob was one of the few whites who commanded the unqualified respect of everyone in the organization. He was a damned good man. No one questioned his courage or commitment. On occasions too numerous to recall, he had put it all on the line. The brutal beatings he had endured at the hands of irate whites while participating in early SNCC marches in Mississippi and Alabama were legend. Like Bob Moses, he was a special SNCC person.

Bob's petition was a stark reminder that placed in sharp focus the heritage of We Shall Overcome, the integrated Freedom Rides, the early voter-registration campaigns, the hope of the Mississippi Summer Project and the despair of Atlantic City. The members of the Central Committee not only had to decide on the future membership of Bob Zellner, they were also being asked to pass judgment on the history of SNCC.

Bob appeared before the Central Committee in late May, 1967.

The Committee, which was chaired by Rap, had been in session all day. Discussion of the petition had been put off until there was nothing else to do but confront it. It was obvious when Bob entered the room that most of the Committee members were very uncomfortable.

The first questions were about the proposed project. Because Bob was a SNCC person, the particulars were just as they should have been. He knew SNCC procedures as well as anyone in the organization, better than most. Rap had been a member of SNCC for two years, five years less than Bob.

Taking his time, Bob explained that he wanted to do the same thing with poor whites as SNCC had been doing with poor blacks. He agreed that he should work with whites. He also agreed that some time in the future coalitions between poor blacks and whites *might* be possible: the classic SNCC position.

After about fifteen minutes, the discussion centered on the crucial issue: Bob's demand that he and his wife be permitted to operate the project as SNCC members. When it was obvious that the time had come for the Committee to make a decision, Rap suggested that Bob leave the room. Before he left, Bob was asked if he would like to make a statement.

"Yes, I would," he replied.

"I respect black people who will stand up for their rights. In the same position, I would expect anybody to respect me for standing up for my rights on a matter of principle.

"I have to take the position that it's either all or nothing. I will not accept any sort of restrictions or special categories because of race. We do not expect other people to do that in this country and I will not accept it for myself."

"I would like to know if you would be willing to have your

status maintained and be excluded from meetings?" asked one of the Committee members.

"My decision is that I am either a SNCC staff member or I am not," replied Bob, who then left the room so the Committee could make a decision.

Ralph Featherstone, new to his position as program director but an old SNCC hand, suggested a possible alternative to the either/or decision before the group.

"I would probably be willing to have the cat on the staff, so long as his voting power were only directed at the policies in the white community," he suggested.

At the same time he was making this suggestion, Ralph's mind was telling him that it wouldn't work. "He, of course, is not willing to accept that position. I don't think we could set up that kind of position anyway because how could a cat listen to discussions and all that type of stuff . . . I guess the only thing we can do is to tell Bob that he is not on staff but that we support him."

"I think to do anything else would be skirting the issue," Stanley Wise pointed out.

Jim Forman, who hadn't said anything, was obviously being torn by conflicting emotions. On the one hand, he was one of the principal architects of Black Power and black independence. On the other, he was closer than anyone on the Committee to the SNCC that Bob Zellner had helped to build and sustain. When he spoke, his words conveyed his anguish.

"I think we are confusing something," he began, "Bob is my best friend. . . ."

When someone interrupted to remind him that he'd said the same thing about Fay Bellamy a couple of days before, Jim exploded.

"That's right, goddammit! I have two best friends. The point is that the issue is not the emotionalism involved, because I know that if I were he, I would be emotional, too. If I had put in that much energy in helping to build this organization in which people are now operating, I would be very emotional. . . . We have to understand that." But though he empathized with Bob, Jim remained committed to the black-only direction in which SNCC was moving.

After much discussion the Committee reached a unanimous decision: Bob was to be fired and all official relations between him and SNCC were to be severed. The Committee decided, however, to provide his project in New Orleans with whatever materials he might request.

After the decision was made, there was concern among some of the Committee members about the manner in which Bob should be notified. They wanted to soften the blow—for Bob and for themselves. Someone suggested that he be informed by letter. Jim Forman disagreed.

"We ought to have enough guts to tell him what we voted on and not try to skirt the issue in terms of a policy that gets us off the hook. We have to tell him just the way it is."

Forman's suggestion was reluctantly accepted and Bob was asked to return to the room. When he arrived, Rap told him of the Committee's decision. Despite the deep hurt in his eyes, Bob remained calm.

"I think it was a mistake, but that is among us. As far as the press is concerned, I don't have anything to say to them about this. I shall continue to operate."

After requesting a copy of that section of tape containing his comments, Bob Zellner left the room.

If SNCC hadn't been involved in a state of perpetual crisis,

Bob Zellner's firing would probably have become the source of a major argument. We were so harried, however, that most of us hardly had time to do more than shake our heads and wish that things had turned out differently. Less than three weeks later, Stokely and several other SNCC members got embroiled in a frightening confrontation in Prattville, Alabama.

It began on Sunday, June 11, when Stokely addressed a group of blacks in a Prattville churchyard. Prattville is located in Autauga County, which is adjacent to Lowndes County. Stokely was urging the people to launch a voter-registration campaign when a white police officer drove up and proceeded to harass him. Because Stokely refused to back down and end his address, the officer arrested him.

Prattville's blacks responded to the arrest by demonstrating in front of the city jail. They demanded that Stokely, who had been charged with disorderly conduct and disturbing the peace, be freed. The police refused to honor this demand. Soon after sundown, guns were brought into play.

Whites shot into several homes during the night, including one that was occupied by a group of blacks who were trying to decide what they might do to get Stokely released. Everyone in the house was forced to lie on the floor to avoid being killed by the torrent of bullets raining through the walls and windows.

Stanley Wise, who had recently been elected to SNCC's executive secretary post, was one of the forty persons trapped in the house. He managed to call the Atlanta SNCC office several times during the night beseeching us each time to send help. He was convinced that they would all be killed unless help arrived. While Stanley was yelling into the phone, I could hear gunshots in the background. We were frustrated because we could maintain our connection with Stanley no more than a few seconds each time he called; a telephone operator in Prattville kept cutting us off.

During one of his calls, Stanley told us that he had heard that Stokely had been lynched. When we called the local jail in order to get the truth, the police wouldn't even admit that Stokely had been arrested.

Frantic, we responded by calling the FBI and the Justice Department. We knew that they didn't really care what happened to Stokely, but that was the best we could do under the circumstances. There was no way for us to get to Prattville before morning.

Near daybreak, a contingent of Prattville police officers and national guardsmen cordoned off a large section of the black community and began to move in with dogs. They were ostensibly searching for snipers. The house in which Stanley and the others were pinned was quickly surrounded and its occupants ordered outside. Stanley and two Alabama SNCC volunteers, Theophas Smith and Ulysses Nunley, were arrested and charged with inciting to riot.

A few hours later the Alabama State Patrol moved into Prattville and took over. Roving through the black community in armed packs, they entered several homes and ransacked them in an alleged search for weapons. John Hulett, chairman of the Lowndes County Freedom party, was one of many persons beaten by the troopers.

At another time in its history, SNCC would have flooded Prattville with organizers prepared to break the will of the white community. Unfortunately, we were no longer in a position to launch such a campaign. We didn't have the staff, the money or the support from outside sources. Being unable to do anything else, the organization authorized Rap to release the following statement at a Monday morning press conference in Atlanta:

> Our course of action has been set. We will no longer sit back and let black people be killed by murderers who hide behind sheets or behind the badge of the law. It is clear that the law cannot and will not protect black people. This is no accident. The racist attitude completely dominates their relationship to the black community and it is blatantly exemplified by their actions. We recognize and accept yesterday's action by racist white America as a declaration of war. We feel that this is a part of America's Gestapo tactics to destroy SNCC and to commit genocide against black people. We are calling for full retaliation from the black community across America. We blame Lyndon Johnson. We extend a call for black brothers now serving in Vietnam to come home to the defense of their mothers and families. This is their fight. We say to brothers in the armed forces: If you can die defending your motherland, you can die defending your mother. It appears that Alabama has been chosen as the starting battleground for America's race war.

Some undoubtedly dismissed Rap's call to arms as mere political rhetoric. We were dead serious, however, because we were convinced that the government had already decided to eliminate blacks by genocide. The events in Prattville were considered part of a sinister plot being carried out on a national level. We didn't believe these things simply because we hated the government.

SNCC members were being systematically harassed in every section of the nation. We were being followed, our phones were tapped and we had only to read the newspapers to confirm what we had suspected for a long time: the organization had been infiltrated by an informer, maybe several. On numerous occasions, decisions made in private meetings appeared in the newspapers before we could pass them on to members of the organization.

Less than six weeks after we managed to get Stokely, Stanley and the others released from jail in Prattville, Rap was wounded in a Cambridge, Maryland, shoot-out. The shoot-out occurred after Rap delivered one of his typically militant speeches at a rally sponsored by the Cambridge Nonviolent Action Committee. Rap was invited to town by Gloria Dandridge (formerly Gloria Richardson) and he spoke at a rundown black elementary school.

After the rally, he was walking a local girl home when a group of Cambridge police officers began shooting into a nearby crowd of blacks. Before he could escape, Rap was hit in the head by a shotgun pellet. Luckily, he wasn't seriously injured. And because it was dark, he managed to get away before the police could move in. After getting his wound treated, Rap left Cambridge and returned to Washington, D.C.

The next day, the attorney general of Maryland issued a warrant for Rap's arrest, charging him with inciting the people to burn a Cambridge elementary school, the same one at which he had spoken the previous night. Later that day, the federal government issued a warrant for Rap's arrest. This second warrant charged him with leaving the state of Maryland to avoid arrest on the first charge. The federal government's warrant, of course, made Rap a fugitive from justice.

Shortly after he heard about the warrants, Rap got in touch with his lawyer, William Kunstler. Kunstler subsequently contacted FBI officials and arranged for Rap to turn himself in to New York City officials the following day. There was no way for any of us to know at the time, however, that the FBI's officials had no intention of permitting him to go free.

When Rap showed up at Washington's National Airport the following day to catch his plane to New York, where he was to turn himself in, he was arrested by a group of police officers who promptly handed him over to the FBI. Although fully aware of Rap's intentions, the federal government chose to arrest him in order to make it appear that he was trying to run away.

During the next few weeks, the FBI and several police groups handled Rap as if he were a football. He was arbitrarily shipped from city to city, jail to jail. Half the time we didn't know where they had him or what new charge they had lodged against him. Rap's lawyers finally managed to get him released on September 18. The government put a major restriction on him before agreeing to let him go, however: he could not leave the eleven counties of the Southern District of New York except to consult with one of his attorneys.

The order had a devastating effect on SNCC. Rap was the chairman of the organization and we counted on him to keep a heavy speaking schedule. These speaking engagements were used to explain our Black Power concepts. More important, they were a prime source of badly needed funds. With Rap under virtual house arrest, SNCC was sorely hampered.

Rap probably would have gotten a great deal of support from those sensitive to civil liberties issues if it hadn't been for the position taken by SNCC in relation to the Arab-Israeli crisis. In the June–July issue of the SNCC *Newsletter*, we had run an article called "The Palestine Problem." There was a great deal of confusion among members of the New Left at the time as a result of the Six-Day War. Many radicals, blacks as well as whites, didn't know anything about the situation. Our article was an attempt to present a coherent analysis.

The editorial on page two of the *Newsletter* explained that the article was designed to answer three questions: "What are the reasons for this prolonged conflict and permanent state of war which exists between the Arab nations and Israel? Why have the hostilities continued? What is the root of the problem?"

The article contended that Zionist imperialists, who'd been active in the area of Palestine as far back as 1897, were primarily responsible for the whole affair. It also said that the United States Government staunchly supported Zionism for neo-Colonial reasons: Israel served as a stepping-stone for the United States Government into Africa. The article very clearly stated that SNCC's members did not consider Zionist imperialism and Judaism to be the same.

The article precipitated cries of outrage from every section of the nation. Attacks came from admitted Zionists, old friends, old enemies and complete strangers.

Folksinger Theodore Bikel, who called the article "obscene," sent us a letter of resignation. In the letter, which he first released to the press, Bikel charged that we had "spit on the graves of Schwerner and Goodman, who died for the concept of brotherhood which you are now covering with shame." Aside from his faulty logic, the thing that most surprised us about Bikel's letter was his belief that he could resign from SNCC. If he had been a member of the organization for five years—as he claimed—none of us knew anything about it.

Author Harry Golden, who edited *The Carolina Israelite,* also lambasted us in the press. He claimed that he, too, was resigning from SNCC after ten years of membership. He was doing this, he said, because of our "increasing use of anti-Semitism" and our echoing of ideas "found in the Ku Klux Klan and the American Nazi Party." As was the case with Bikel, none of us knew anything about Harry Golden's SNCC membership.

I don't wish to create the impression that all our critics were white. Such was not the case. As a matter of fact, some of our most vehement critics were Negroes. Our old friend Bayard Rustin joined A. Philip Randolph and issued a declaration in which they said that they were "appalled and distressed by the anti-Semitic article."

The most important result of the article was that it led to the end of all significant financial support from the white community. The little money we were still receiving from white sources just stopped coming in. Rather than breaking our will, this made us more convinced than ever that we were correct when we accused the majority of America's whites of being racists.

CHAPTER 16

"Soon, we will not cry."

On the last day of the first week of October, 1967, Ruby Doris died. She had been severely ill for ten months. I think she died from an intestinal disease, but I'm not certain. The funeral, which was held in Atlanta, tore me up. She'd meant so much to me. I'd respected her from our first meeting in Rock Hill; she was one of the persons who inspired me to make the movement my life. I had loved her. If she had taken it easy, if she hadn't stayed with the SNCC life until it completely broke her health. . . .

The September–October issue of the SNCC *Newsletter* contained an item that conveyed our utter frustration. It was initially a letter written by Fay Bellamy when she heard about the death of two SNCC members in West Point, Mississippi. The friend to whom she had written the letter returned it to her in the form of a poem. It was titled "Soon, We Will Not Cry."

Soon, We Will Not Cry
By Fay Bellamy

John Buffington just called from West Point, Mississippi, and said George Bess and Henry McFarland had been in a car "accident" and both had drowned.

John thinks they were forced off the road. He said they talked to the Sheriff of Clay County and the Sheriff states they weren't driving fast and there were very few skid marks. They were crossing a bridge and the car went off and turned upside down.

Oh Baby, I felt—and still feel—that death is going to be a big star in the life of black people. When people were killed in Detroit and Newark, and all the other places, I felt pain each time a so-called statistic was added. The pain is deeper here because I knew George, and maybe one day I might have known Henry.

Ricks left the house when I told him about it. Stanley went to the office. Both cried.

I guess what I'm trying to write about is the pain I feel at this moment. Can one write about pain? What I'm also attempting to ask is how does one get used to it?

How many people will have to die before we can make it a two-way street? I'm afraid of war, never having known it, but I'm even more afraid of how many of our people have to die.

I would much rather us die fighting to defend ourselves, since we die all the time anyway.

I want to cry but am not able to do so.
With each death we cry a little less.
Soon, we will not cry at all.

CHAPTER 17

The Orangeburg Massacre

I ARRIVED IN Orangeburg, South Carolina, in early October, 1967. The weather was nice, I was rested for the first time in years and felt confident that I could get some good things going. My first objective was organizing students on black college campuses, especially those at South Carolina State. By working with the students, I believed I could develop a movement focusing attention on the problems of poor blacks in South Carolina. I also hoped that my efforts might encourage some of SNCC's members to return to the South and begin organizing again.

For the first few weeks, things went just as I'd expected. Officials on the campuses were suspicious, but not openly hostile. Many of the students I approached were interested, but afraid of administrative sanctions.

"Things haven't changed much in South Carolina," I told Sandy one evening after a meeting with a group of students at South Carolina State.

"These kids are being controlled by the same tactics that were used by the administration at Voorhees when I was a student."

By mid-October, I managed to establish a working relationship with a group of students on South Carolina State's campus. They belonged to an organization called The Black Awareness Coordinating Committee (BACC). Although most of the students in the group were politically moderate, they were all very interested in Black Power.

In mid-October, Sandy and I were joined by John Batiste and Bill Ballon. John was a veteran SNCC member from South Carolina. And Bill was a volunteer from the West Coast. We worked together well and soon began to visit students on other campuses. We also began to make contact with people living in Columbia and other surrounding towns.

We'd been working together for a couple of weeks when I became aware that we were being watched. I learned about this from a small article printed in one of the local white newspapers.

The article claimed that a "group of long-haired black militants" was traveling around the state trying to stir up trouble among Negroes. There was no question in my mind about the identity of those "long-haired black militants."

"We have got to be very careful," I told John and Bill. "South Carolina is not like a lot of states. Some of these crackers will just as soon kill you as say 'good morning.'"

"You don't have to tell me that," John replied, "I grew up here, too."

Although we tried to keep in touch with the National Office in Atlanta, it was very difficult. There just wasn't any organization in Atlanta. Moreover, the emphasis was not on organizing. If we hadn't been able to scrounge support from people in the area where we were working, we would have starved.

The hours of work that John, Bill and I put in proved fruitful. BACC gained in prestige with each passing day. Although it didn't have many members, by late December it was one of the most important student groups on campus. The NAACP, which had a large chapter on campus, brought in several prominent speakers in order to combat BACC's influence, but this tactic backfired. Most of the speakers, instead of parroting the NAACP's moderate line, challenged the students to be more militant.

Sandy and I were having numerous difficulties during this period. Although I had more time to spend with her than before, we still couldn't get it together. She didn't have the same fervor for the movement as I and couldn't understand why I was so obsessed with it. She was still very insecure and continually worried that I would leave her.

We talked about her insecurities on several occasions. She was convinced that they would go away if we got married. Although I disagreed, I told her that I would marry her if she thought that would help. In mid-January, we drove from Orangeburg to Atlanta to "get hitched."

Though I agreed to do it, I really didn't want to get married. This doesn't mean that I didn't love Sandy. I did. It's just that I consider marriage to be a stupid, outdated institution. If people love each other, that should be enough. I see no need to give the state money in order to make living together "legal." What two people in love feel for each other has, in my opinion, nothing to do with the state. It's between them and them alone.

I was so much against the idea of marriage that I tried to get Willie Ricks to go downtown and get the license for me. When he refused, I had to do it myself.

On the twenty-second or twenty-third of January, 1967, I forget which, we were married. Stokely had agreed to be my best man but got tied up at the last minute and couldn't make it. That was okay. Dr. King, who had suggested many times before that we get married, performed the service at his church.

As I had suspected, marriage didn't eliminate Sandy's insecurities. After a couple of weeks, things were just as bad between us as they had ever been. As a result, Sandy decided to leave me for two or three weeks. I hated to see her go, but realized that it was probably for the best.

We parted amicably. We agreed to think about our problems while separated and try to come up with some ideas about how we could save our relationship. When she returned, I wasn't there to meet her. Unforeseen events demanded that I be elsewhere—in prison.

BACC's members were approached on several occasions during the fall and winter by students concerned about a segregated bowling alley in downtown Orangeburg. The bowling alley, which was owned by a man named Harry Floyd, was the only alley in town. The students wanted it integrated. They had been protesting Floyd's exclusionary policy for more than two years, during which time the bowling alley had become a hated symbol of discrimination.

BACC's members repeatedly refused to get involved in the bowling alley issue. They believed, and I supported them, that integration was an irrelevant issue. Despite BACC's position, a group of insistent students refused to let the issue die.

I was out of town when the first demonstration was held at the bowling alley on the night of February 5. Fifty students from South Carolina State confronted Floyd and demanded that they be permitted to bowl. Floyd denied their demands and insisted that they be arrested. Orangeburg's police chief, Roger E. Poston, refused to arrest the students. He said they hadn't committed any crime. Chief Poston did insist, however, that the students leave the bowling alley.

I returned late the following day. As was my usual practice, I took a quick turn around town to see if everything was in order—a precautionary measure I had developed while serving as project director in Holly Springs. When I got to the South Carolina State campus, I was approached by two co-eds who were members of BACC.

"Several people have been arrested down at the bowling alley," one of the girls said.

"You've got to go back with us," they insisted. "We don't know what's going to happen to them."

"Why were they arrested?"

"I don't know for certain," one of the co-eds replied. "They went into the bowling alley. And the next thing I knew the cops were bringing them out and putting them in cars."

"Okay, let's go," I said.

We moved toward the bowling alley along with a swelling stream of students headed for the same place. Many of the students were picking up rocks and bottles as they walked. Although the majority were from South Carolina State, some of the students were from Claflin, a small black college located across the street from State.

While walking, I talked with several students, trying to find out what had happened. One of the students who had been at the bowling alley earlier had returned to the campus to get help. He filled me in.

"Everything was cool when we first got there," he said. "John Stroman [a student leader from State] was in charge. They must have known we were coming because the doors of the bowling alley were locked. For some reason, though, they opened the doors.

"A bunch of us went inside and tried to bowl, but they wouldn't let us. After a few minutes, a police officer told Stroman that those of us who didn't want to get arrested should leave. He said that it would only take a couple of arrests to make a court test of the bowling alley's segregated practices.

"That's when most of us left. Only about fifteen people stayed inside. They were arrested. And the cops brought them out and put them into police cars.

"Everything was cool until the cops rushed into the crowd of students outside in the parking lot and arrested some cat. That's when I headed back to the campus for help."

When we arrived at the bowling alley, things were relatively quiet. Although there were three or four hundred students milling around in the parking lot adjacent to the bowling alley, the police weren't bothering them. I saw a member of BACC and walked over to talk with him. He told me that a compromise had been worked out. The police had agreed to release the arrested students in the custody of one of the deans from State. And the students had agreed to end the demonstration.

Everyone was preparing to head back to campus when two large fire trucks pulled into the parking lot. The students, who thought that they had been tricked, were infuriated.

"Hey, man. Where's the fire?" yelled someone.

"The mothafuckas are trying to get away with some shit," someone else yelled.

The students were angered by the fire trucks because they were reminded of the experience they had had in 1963 when Orangeburg officials used fire-fighting apparatus to break up student demonstrations. Several students had been knocked down and swept under cars by the force of the water. Other students, who had been arrested, had been hosed down while they were locked in an outside pen surrounded by a high wire fence. The weather had been cold and several students almost contracted pneumonia.

As soon as the police realized that the fire trucks were angering the students, they asked the firemen to leave. It was too late. By this time a small group of students was trying to enter the locked doors of the bowling alley. A contingent of police officers rushed forward to stop them and a pushing-shoving melee ensued.

When one of the students kicked in the small window next to the main doors of the bowling alley, the police, who had been relatively civil, seemed to lose all composure. They raised their long nightsticks and began to flail away at the surprised young men and women. It didn't matter who they hit, male, female or innocent bystander. Those in the front of the crowd were beaten quite severely; there was no place for them to run.

On the way back to campus, the fleeing students took out their frustration on white-owned stores along the route. Display windows were broken, trash cans turned over and mannequins upended. Rocks were scraped along the sides of several parked cars and some radio aerials were broken. There was no looting.

Subsequent estimates of damage indicated that the East End Motor Company was hit hardest. Eighteen hundred dollars' worth of damage was done to its property, four times that of any other business.

Later that night, there was a large meeting in one of the campus auditoriums. The atmosphere was tense and filled with emotion. Several students and faculty members spoke. Everyone wanted to do something about the bowling alley and other conditions in the town, but nobody seemed sure what was appropriate. At one point, I proposed that they use their bodies to block all traffic along College Avenue, a main street adjacent to the campus. This relatively moderate suggestion was dismissed as being too radical.

When I heard that a group of students and instructors were holding a meeting at the home of Professor Roland Haynes, I decided to attend. Professor Haynes was the head of the Department of Psychology at State and faculty advisor to the student NAACP chapter. I sat in the meeting for a short time. For the most part, they were just as confused as the students.

After much discussion, it was decided that a demonstration should be held in downtown Orangeburg the following day. Someone wondered what we should do if city officials refused to give us a permit. It was decided that the march would take place—with or without a permit.

It was after twelve when I left Professor Haynes's house and headed home. I was very tired. When I reached home, a state trooper was parked across the street. He watched me, but didn't say anything. He remained there all night.

When I arrived on campus the next morning, plans for the downtown demonstration had been scrapped. Someone had gotten in touch with the mayor and other city officials and they had insisted that it was too dangerous. An afternoon meeting had been scheduled between the students and the mayor in a campus auditorium.

The first thing the mayor did was disclaim all responsibility for the actions of the police. He then proceeded to expose the fact that he had almost no understanding of the students or their problems. The longer he talked, the more restive the students became. By the time he finished his address, they were on the verge of booing him out of the auditorium.

BACC's president, Wayne Curtis, asked several questions of the mayor and the assistants he had brought with him. Most of his interrogation dealt with racial discrimination in Orangeburg. The mayor and his assistants were both surprised and embarrassed by the questions. Their answers consisted mostly of stammers and vague excuses, which only angered the students.

Realizing that they were exposing ignorance and prejudice, the mayor and his assistants relinquished the stage. I didn't find out until much later that they thought that Wayne was me and that I had been deliberately trying to make them look foolish.

Later in the day, students from South Carolina State and Claflin got together and drew up a list of moderate demands. The most important were:

1. Close down the All Star Bowling Lanes immediately and request the management to change his policy of racial discrimination before opening.

2. Police brutality—The action taken by the . . . officers was uncalled for, especially the beating of young ladies.

3. Immediate suspension, pending investigation, of the officer who fired a shot unnecessarily into the State College Campus.*

4. The establishment by the Mayor of an Orangeburg Human Relations Committee of a biracial nature, with the recommendation that each community select their own representation.

5. A request should be made for a public statement of intent from the Orangeburg Medical Association as to its determination to serve all persons on an equal basis regardless of race, religion or creed.

* A white police officer drove up on campus after the Tuesday night fracas and fired his gun into the air because rocks were allegedly thrown at his car.

6. Formulate or integrate a fair employment commission in the city of Orangeburg.

7. Change the dogmatic attitude of the office of personnel at the Health Department and the segregated practices used there.

8. Extend the city limits of Orangeburg so as to benefit more than one segment of the community.

9. Give constructive leadership toward encouraging the Orangeburg Regional Hospital to accept the Medicare Program.

10. Eliminate discrimination in public services, especially in doctors' offices.

11. The integration of drive-in theaters.

12. Fulfill all stipulations of the 1964 Civil Rights Act by leading the community so that it will serve all people.

These demands were presented to city officials at City Hall on Wednesday afternoon, at 4:30 P.M. Although the list was not accompanied by a deadline, it was obvious from their attitude that the officials didn't plan any immediate action. This increased the level of anger and frustration among the students, who were convinced that they were being given a runaround.

Soon after dark they began to throw bottles. Because it was considered unsafe to venture off campus, most tossed the bottles and rocks at cars driven by whites along College Avenue.

At one point during the night, three students were shot-gunned while walking down a street near the campus. They were shot by a white man who later told police that he believed they were preparing to attack his house. This incident served to heighten tension and increase anger. Luckily, the students weren't seriously injured.

Shortly after 11 P.M., a car containing two white men came careening through State's campus. One of the men reached out of a window with a pistol and began firing it wildly at the students. I dived for cover behind a clump of bushes at the sound of the first shot. The driver rode into a dead-end street where his car was bombarded by rocks and bottles, but he turned around and sped off campus.

We all believed that the police, who had set up roadblocks on the streets surrounding the campus, had let the car through with the hope that some of us would be killed.

"Those mothafuckas never could have gotten through the roadblocks if the police hadn't let them," a student declared.

As I was about to return home, I was informed that there were four highway-patrol cars parked in front of my house—with two patrolmen in each car. It sounded like a setup to me. I asked one of the campus security guards if he would accompany me but he refused. "Do you think I'm crazy?" the guard asked.

I spent that night in a dormitory room on campus.

On Thursday morning, South Carolina State's acting president, Maceo Nance, had a memorandum circulated among the students. It said what you would expect a college president's memorandum to say under such circumstances.

". . . your personal safety is in jeopardy, and we are requesting that all students remain on campus and refrain from throwing brickbats and bottles as was the case last night. The shooting last evening bears out the danger involved in this kind of violence and destruction.

". . . Until some semblance of order is restored, students are requested to remain in the interior of the campus."

Shortly before noon, I returned to my house in order to wash up and change clothes. Although I lived only one block from campus, I didn't feel safe. Several members of BACC accompanied me. I didn't know what we might encounter.

Fortunately, the highway patrolmen were gone. There were several newspaper reporters waiting for me instead. Like the governor, the mayor and seemingly every other white official in the state, the reporters were convinced that I was responsible for everything.

"Everybody is looking for a scapegoat," I told them.

They wanted to know what was going to happen next, I told them that I had no idea. They wanted to know if Stokely was coming to town; it was obvious that they were hoping that he wouldn't.

"Stokely probably won't come to town, unless I'm incarcerated," I told them. I was hoping that this information would get back to the police. I was certain that they intended to arrest me, and this comment was designed to make them change their minds.

I returned to the campus immediately after the reporters left. I felt relatively safe there. I did not believe that the police would attempt to get me as long as I remained at the school. Although President Nance had instructed the dean to keep all nonstudents out of the dormitory, I planned to continue sleeping in the dormitory until the crisis was over.

The tension that suffused the campus during Thursday was unbelievable. Everybody was talking about the beatings, the shootings, the bowling alley and the cops. Reports from the surrounding black community indicated that there were hundreds of state troopers and national guardsmen in town. There were also reports that white citizens were in the streets with guns.

Everywhere I went I heard people say, "Something's gonna happen. Something's gonna happen. Something's gonna happen."

At approximately 8 P.M., I went to one of the dormitories for a nap. I knew that it was going to be a long night and I wanted to get some rest. Because I was extremely tense, I had trouble getting to sleep. Somewhere around 10 P.M. there was a loud commotion out in the hall.

"They're doing it. They're doing it. You all better come see," someone was yelling.

Jumping from the bed and stumbling over the clutter of shoes and books in the middle of the floor, I made my way to the door.

"What's going on? Where is everyone going?" I yelled to one of the students as he rushed past.

"I don't know! Something's going on down by the edge of the campus," he called out.

Hurrying back into the room, I grabbed my pants and shoes. Dressing quickly, I headed for the door.

"Just some more bottle throwing," I muttered to myself and rushed out into the cool night air.

When I arrived at the scene of the action, my heart began to pound, hard and fast. The students had been driven back toward the center of the campus where they had built a huge bonfire in the middle of Watson Street right off College Avenue. Hundreds of national guardsmen, state troopers and police officers were in a line facing the students on the other side of the fire. There was probably no more than fifty yards between them. The flames of the fire highlighted the silvery bayonets of the national guardsmen's rifles.

"What's going on?" I asked as I walked slowly toward the front of the crowd. No one took the time to answer. The students were engrossed in the fire and the armed men stationed on the other side of it. Most of them appeared to be calm, confident that nothing would happen because they were on *their* campus.

Then I saw Henry Smith, a tall lanky sophomore, standing in the front of the crowd. Henry stepped forward to place something on the fire, when the cops started shooting. Amid terrified screams, I watched him spin and crumple to the ground. Bullets seemed to be coming from all directions. The sound of the pistols and shotgun blasts was deafening.

I turned and dived head first for the ground. Something hit me in my left shoulder. It felt like a power-driven sledge-hammer. I landed with a heavy thud. The air was knocked out of my lungs but my mind was working fast.

I was certain that they had shot Henry thinking he was me. He looked just like me. I began to crawl, trying to get out of the line of fire. After the students' first cries of pain and horror most were quiet. Only the wounded made sounds. They were either moaning or begging for help.

Those who hadn't been hit quickly crawled away. Those of us who had been hit moved slowly. Others did not move at all.

Although my entire left side was getting numb, I was frightened for my life. I reached a brother and tried to help him. He could hardly move and the blood was spurting out of several wounds in his back.

I pulled him for about ten yards to a small bush—out of the line of fire. I had to leave him there because I was just not strong enough to drag him any farther. The guns were sounding louder now and the cops seemed to be advancing. I knew that if they caught me, they would kill me.

"You'll be okay here," I tried to explain to the fellow. "I've gotta go, man!"

He could not answer me, but I think he understood. I crawled as fast as I could through the dark. Ahead was a slight depression in the ground. I could feel it in the cool grass. For a few moments, I rested in the depression, which shielded me from the bullets. Then I crawled another fifteen yards and came to one of the cross-campus streets. I rolled painfully into the gutter, where I was able to slide along the street side a bit faster. The bullets were still coming and I was terribly afraid that I would lose consciousness before I could escape.

When I had crawled far enough away from the gunfire, I got up and ran to the other side of the street. Pausing for a moment, I tried to get my bearings. There was a small tree about twenty feet away; I headed for it. Halfway there I stumbled. But, I gathered all my remaining strength and headed for that tree, for I knew that just beyond it was a dormitory. And when I did finally make it, I collapsed two steps inside the door.

"Cleve. Oh, no!" I heard someone scream, a girl. "Are you hurt bad? Where? Tell me where else are you hurt, Cleve." She had been bandaging a bleeding student who was lying next to me. She did not stop and help me until she was finished. I do not believe we would have made it without people like her—those who rolled up their sleeves and came to our aid in the bloodiest event many of them would ever see.

"I think I'm gonna be all right," I answered weakly.

CHAPTER 18

Prisonbound

AFTER RESTING IN the girls' dormitory for a few minutes, I got up and headed for the infirmary. I felt better and wanted to help those with more serious injuries. Although I had expected things to be bad, it was worse than I had imagined. Blood was everywhere, on the floor, the walls, the chairs; and everyone was working with the wounded.

Most of the forty to fifty students in the infirmary were quiet, though some were weeping softly. One of the students, I think he was the quarterback on State's football team, was paralyzed. He had been shot in his spine.

There was only one nurse on duty. And she couldn't administer aid to all those who needed it. She put in a call to the athletic director and the coaches. They did the best they could, but many of the injuries required the attention of a doctor. Finally, the nurse suggested that all the injured be taken to the city hospital.

Although I was worried about my wound—it had begun to send waves of excruciating pain through my entire body—I didn't want to go to the hospital. I knew that the police would be there and I was afraid that they would kill me. I asked the nurse to keep me in the infirmary and treat me herself, but she refused. She said that I needed the attention of a doctor.

A college instructor finally convinced me that I should go to the hospital. He promised to remain with me and to keep the police from hurting me.

We were transported to the hospital by students who had brought their cars to the infirmary. The police had surrounded the campus and they weren't allowing any ambulances through. When we got to the hospital, the police refused to permit the instructor to accompany me into the emergency room.

"If you're not wounded, get the hell away from here," a cigar-chewing officer said. He was holding a double-barreled shotgun and it was obvious that he would gladly use it if we gave him an excuse.

"I'll be okay," I told the bewildered instructor.

I thought at the time that I could get assistance and leave the hospital before the police recognized me. There was a great deal of confusion. I attempted to hide by sitting behind two large chairs.

While waiting to see the doctor, I decided to call my parents. I wanted to assure them that I was okay before they were frightened by unfounded news reports. Although they were alarmed when I told them what had happened, they remained calm.

"Are you hurt? Where are you? Tell us the truth."

"I got shot in the arm, but I'm okay," I said.

I could tell from their voices that they didn't believe me. My father insisted, despite my reassurances, that he was coming to Orangeburg.

"Where are you right now?" he asked.

When I got back to the room where they were keeping us, the police were bringing in a pregnant black woman. Her hair was tousled and she was in hysterics. I later found out that she was a married student from South Carolina State who was arrested while transporting wounded students to the hospital.

"I've been beaten. I've been beaten," she repeated over and over again.

Handling her very roughly, the police hauled her atop a table. After lying there for a few minutes, she screamed, grabbed her stomach and fell to the floor. Five or six brothers jumped up and rushed over to help her. They moved about four steps before six armed police officers, all dressed in riot gear, jumped up and told them to "Get back over there. Don't let's get started again."

The woman, whose name I found out later was Louise Kelly Cawley, lay on the floor writhing in pain until a couple of orderlies found the time to place her back on the table. She lost her baby a week later.

While we sat there helplessly watching Mrs. Cawley, we got word that one student was dead and that several others were expected to die.

I had been in the waiting room for about a half hour when a black police officer came through. Our eyes met for a moment; there was a flicker of recognition. I knew immediately that I was in big trouble. Without saying a word, the officer turned and left the room.

I wanted to get up and run, but I knew that escape was impossible. Cops were everywhere. In less that five minutes, the black officer returned. He was accompanied by two white officers. They beckoned to me. I acted as if I hadn't seen them.

"Sellers. Come with us."

They took me immediately to a doctor. He examined my throbbing arm and told me that I was lucky. "You're not hurt bad." After binding my arm, the doctor gave me a shot. I was then taken back to the waiting room. Before I could regain my seat, a white man in shirt sleeves walked up to me and said, "Come on!"

"Who are you?" I asked.

"I'm the sheriff. Let's go."

"Am I under arrest?"

"Don't give me any trouble. Let's go," he commanded.

As they led me from the hospital, I called to every student I passed.

"I'm with the sheriff!"

"I'm with the sheriff!"

"The sheriff's got me!"

"If they kill me, you know who's responsible!"

"The sheriff's got me!"

I thought that it was all over at that point. Visions of the burned-out car in which Chaney, Goodman and Schwerner were riding when they were last seen flashed through my head. I thought of their broken, bullet-riddled bodies. My legs were rubbery, but I was determined not to show fear. To my surprise, they didn't take me out into the woods. They took me instead to the courthouse.

I was arraigned that night. While standing drowsy and filled with pain before the judge, I heard the charges against me: "arson, inciting to riot, assault and battery with intent to kill, destruction of personal property, damaging real property, housebreaking and grand larceny"—seventy-eight years worth of charges.

My bond was set at fifty thousand dollars. At the conclusion of the hearing, I was placed in a car and whisked forty miles away to the state prison, where I was locked in a tiny cell on death row.

During the next few days, I received bits of information: three dead and twenty-seven known wounded. Henry Smith, Delano Middleton and Samuel Hammond, Jr., were the dead students. The shotguns used by the police had been filled with deadly double-aught buckshot (double-aught buckshots are about the size of a .32 caliber bullet).

With only two exceptions, all the dead and wounded were shot in the rear or side. Some were shot while lying on the ground or attempting to crawl away. Many of the wounded were shot in the soles of their feet. In some cases, students were standing as far as one hundred yards away from the wildly firing officers when they were hit.

City and state officials, and particularly Governor Robert McNair, attempted to whitewash the entire affair. They claimed that the police fired in response to "intense sniper fire coming from the campus." The news media accepted and reported this lie as if it were gospel.

SNCC attempted to set the record straight with a two-page news release, but it had little effect. The organization just wasn't equal to the task. Unlike Selma, where SNCC was able to mobilize national support within hours after the attack on the Edmund Pettus Bridge, there was almost no response to Orangeburg. SNCC's elaborate communications apparatus was gone. Moreover, there was no money for my bail.

Prisons are strange. Although the inmates are locked in their cells most of the time and watched like laboratory cultures, they do manage to communicate with each other rather well—through walls and from building to building. I had been in prison for only a few hours when I received a letter from a brother named Bostick. He knew who I was and why I was there.

The letter, which was quite short and written in pencil, raised my sagging spirits. A couple of days later, he sent me a second letter:

"Brother Sellers:

"By being a native of Denmark, I assume you heard about me when you were in high school. If not, I am the boy who was accused of murdering the sheriff of Jasper County. I've been on death row for 6 years. I was fighting for a new trial for about five years and two months before the United States Supreme Court granted me one. . . .

"I have caught a whole lot of hell since I've been here. Six years of nothing but frustration and confusion. I wish like hell that you're able to get out of this place before long. . . ."

In subsequent letters, Bostick told me about the way he had been forced to sign a false confession and the manner in which his case had been railroaded through court. Even so, he was still optimistic that he would be released.

I don't know how he knew so much about me or how he managed to get the letters from his cell to mine. All I know is that every once in a while, I would look up and another letter would be lying on the floor of my cell.

One day, I received a letter from John Batiste. He'd been arrested in Orangeburg while driving my car. The police took him to the police station and held him. After he'd been there several hours, a highway patrolman walked in with a .38 caliber pistol and a box of .38 bullets. John was subsequently charged with "breach of the peace, violation of the curfew and carrying a concealed weapon." He was being held under five thousand dollars' bond in another section of the prison.

My parents visited me several times. They were very worried about my safety, but I assured them that I was all right. One afternoon, my father came to see me.

"How is it going?" he asked when I entered the visitor's room.

"I'm okay."

I felt good seeing him. While we talked, he told me about the trip he had made from Denmark to Orangeburg on the night of the killings. I don't know what it was, but something happened to him that night. Something made him realize that he really wasn't as secure as he had been trying to believe all those years.

"I'm going to buy a new car," he said.

"What kind?"

"A Cadillac! You only live once."

Most of the rest of our conversation was unimportant. We said a lot with our eyes. For the first time since I was a boy in high school, everything was okay between us. He respected me for what I was. And I finally respected him.

"Don't let 'em getcha down," he said to me as the guard led me out of the room and back to my cell.

If SNCC's attorney, Howard Moore, hadn't managed to get my bail reduced, I probably would have remained in prison much longer; SNCC was flat broke. Howard is an excellent lawyer, however, and after only three weeks I was released on a twenty-thousand-dollar bond. One of the stipulations to my release was that I could not under any circumstances go within five miles of Orangeburg's city limits. Although I didn't like this edict, there was nothing I could do about it.

During the time I was in prison, Jim Forman sent George Ware to Orangeburg to take charge of the student movement. Immediately after my release, I had a talk with George in Atlanta. He suggested that I could be most useful by going on a fund-raising tour. I didn't like the idea, but agreed to go.

I traveled from city to city, until mid-March, when I returned to Atlanta. The federal government was finally ready to try me for refusing to be drafted. On March 29, 1968, seven weeks after the Orangeburg assassinations, I went to trial.

Julian Bond and John Lewis appeared as character witnesses for me. Charles Webster, a white man who had formerly served as chaplain at Clemson College, also spoke in my behalf. The jury, which was composed of nine whites and three Negroes, was not moved by their testimony. Nor was it moved by Howard Moore's eloquent plea in my behalf. I was found guilty. The judge instructed me to return for sentencing one month later.

I was not surprised or depressed by the verdict. By that time, I would have been surprised if I hadn't been found guilty. Howard had told me before the trial that my chances did not look very good. He thought I might get a three-year sentence.

"That's what they've been giving people convicted of this charge."

I just nodded. When you're faced with seventy-eight years in prison, three years one way or another doesn't mean very much.

As soon as my trial was finished, I headed for New York. Jim Forman was there and there were some things I wanted to take up with him—the three weeks I spent in prison. We had a big argument.

"Goddammit! What the hell's going on around here?" I asked. "Why the fuck wouldn't you guys get off your lazy asses and get me out of prison sooner than you did?"

Forman insisted that everyone had done everything possible to get me released. He said that it probably would have been arranged sooner, but Rap had violated his probation and had gone to California.

"A lot of the attention which would have normally been directed to the situation in South Carolina was devoted to Rap," said Forman. "He's the chairman of the organization. I'm sorry you got jammed up, but that's the way it goes sometimes."

While in New York, I heard that Stokely was going to conduct a big press conference in Washington, D.C. The press conference, which was to be held on April 5, was going to be devoted primarily to Rap. He had been imprisoned in Louisiana for more than a month and was scheduled to come to trial within a couple of weeks. The press conference had been well publicized and it looked as if all the major newspapers and television stations would be represented.

"Hell, I might as well go down to Washington," I told Forman. "Maybe I can get some exposure down there. If I don't do something, I'm going to end up spending the rest of my life in some South Carolina prison. Nobody knows a goddamn thing about what happened in Orangeburg, including most of the people I met on that fund-raising tour."

I caught a plane out of New York City on April 4, 1968. I was so tired that I could feel the blood coursing through the veins in my temples each time my heart beat. I tried to read a magazine, but couldn't keep my eyes open. I tried to think about what I would say at the press conference, but my mind refused to focus. Finally, I gave up and went to sleep.

I ran into two SCLC officials, Bernard Lafayette and Stony Cook, while making my way through the crowded airport in Washington. They seemed upset, but I was so tired I wasn't really listening to them. They said something about trying to get to Memphis.

"Yeah, Dr. King's down there leading the garbagemen," I replied. "When you see him, tell him that Cleve Sellers said, 'Black Power is the answer.'"

At this point, Stokely walked up. I had called him before leaving New York. His eyes were red and he seemed upset, too.

"What's the matter, Stokely? You look worse than me," I said, chiding him.

"Haven't you heard?"

"Heard what?"

"Dr. King has been assassinated."

CHAPTER 19

The King Is Dead

Dr. King has been assassinated. Dr. King has been killed. Someone shot Dr. King in the head. They got him. Some honkie shot him in the head. Dr. King is dead.

Over and over. These words kept registering and reregistering somewhere between my ears. I tried to relate them to reality, to the good-natured man I'd gotten to know while trudging along the back roads of Mississippi. I didn't want to accept it. He was so full of love, so full of life.

"There must be some mistake," I said to Stokely at one point while we were riding toward the Washington SNCC office.

"No, there was no mistake," he replied. "The dirty motherfuckers got him."

It took me about a half hour to get myself together. It probably would have helped if I could have cried, but I couldn't. I was all cried out. Like everyone else in the crowded office, I had seen too many murders, attended too many funerals, grieved for too many lost and never to be seen again friends.

Rage and numbness dominated my emotions. Sitting alone in

the corner, all I could do was stare blankly at the walls and try to swallow the huge lump in my throat.

Stokely was very upset. He is volatile and tends to have little control over his emotions when he is angry. His eyes reflected pure rage.

"The dirty motherfuckers. The dirty motherfuckers," he kept repeating. "There was no reason. There was absolutely no justification."

Somewhere near 8:30 P.M., Stokely stood in the middle of the paper-cluttered floor and made an announcement. "They took our leader off, so out of respect, we're going to ask all these goddamn stores to close down until he is laid to rest. If Kennedy had been killed, they would do it."

With that, he bolted out the door of the office. I looked at Bill Hall, who was standing on the other side of the room. We headed for the door. "We've gotta stay close and keep him from getting in trouble," I said as we stepped into the street.

"Yeah, he's in a bad way," he replied.

Hurrying down the street, we caught up with Stokely after about a half-block. We were in the heart of Washington's sprawling black ghetto. There were quite a few people on the street. It was obvious from the expressions on their faces that most of them had heard about the assassination.

There were several small groups of people clustered around transistor radios, attempting to get additional news from Memphis, trying to piece the whole story together.

I looked back at one point. There were about twenty people following us. Most of them were members of the Washington SNCC staff.

The first place we stopped was a black-owned barbershop. When the proprietor was asked to close up in remembrance of Dr. King, he readily agreed to do so. We stopped next at a Chinese restaurant. After Stokely spoke with the owner for a few minutes, he too agreed to close. Striding from store to store along the crowded street, Stokely asked manager after manager to close his business. All of them did. The crowd behind us grew larger and larger.

Although the situation was not yet ominous, I was very tense. I wondered whether Dr. King's assassination had been part of a nationwide plan. They may have decided to get all of us at the same time and get it over with, I thought to myself. It makes sense. If they kill Stokely, Rap, Huey Newton and the rest of us, the movement will be thrown into total disarray. Stokely shouldn't be out here where anyone who wants to can easily gun him down.

My thoughts were interrupted by the shouts of a black teenager standing on the periphery of the fast-growing crowd.

"Stokely, you're the one," he screamed.

"Now that Dr. King's dead, we aint' got no way but Stokely's," yelled another.

At one point, we were approached by Reverend Walter Fauntroy, an official with the Washington City Council and one of Dr. King's regional representatives. I could tell from the expression on his face that he was worried about the huge crowd gathering around us. Catching up with Stokely, he voiced his concern about the possibility of trouble.

"Stokely, Stokely, this is not the way to do it. Let's cool it."

Fauntroy, who is much shorter than Stokely, was hanging on to one of his arms. Continuing forward with forceful strides, Stokely seemed oblivious of his weight.

"All we're asking them to do is close the stores," Stokely said.

Fauntroy, who seemed somewhat mollified, let Stokely go. Stokely continued to walk without looking back. The crowd in his wake streamed around Fauntroy and left him standing in the middle of the sidewalk.

By 9:30 P.M., some members of the crowd began calling for retaliation against white-owned business establishments in the neighborhood. Many began to pick up rocks and bottles. When we passed the Republic Theatre, one young boy in the crowd stepped forward and rammed his fist through one of the glass doors. Another youngster stepped through the glassless door frame and emerged with a large bag of popcorn.

Stokely turned around at the sound of the shattering glass and confronted the youth who had broken it.

"This is not the way," he shouted.

During the next half hour or so, the inevitable began to happen. Enraged by the senseless assassination, people in the community began to stream into the streets. They were looking for some way to let out their frustrations. We tried for a while to keep them from moving before they were prepared to deal with the police.

"Take it easy, Brothers! Take it easy, Brothers," we kept repeating. But to no avail.

By 11 P.M., it was obvious that we could do nothing to dissuade the people from reacting. They intended to have some kind of revenge. Our telling them to "Go home so that you won't be slaughtered" had little effect. There were too many of them and too few of us. However, we were able to convince Stokely to leave. He too realized that we were powerless to stop the looting, which spread quickly.

We went to Stokely's apartment to talk and to get ourselves together. His place was located in the heart of the ghetto, near the center of the disturbances. From time to time, we heard windows being broken, fire alarms going off, police sirens, loud yells and gunshots. It sounded as if a war was being fought. Although we had the windows closed, the acrid smell of tear gas and burning wood filled our nostrils.

Though we kept the radio on to get additional news from Memphis, none of the periodic broadcasts helped us much. All we could really learn was that Dr. King had been shot somewhere in the head and that he was dead. Most of the rest of the news was either conjecture or unfounded rumor.

We talked late into the night. Stokely, who was still extremely upset, recalled many experiences he'd had with Dr. King. Although I had known that they were close, his comments indicated that they were much closer than any of us had ever imagined.

They had met for the last time in February, when Dr. King went to Washington to complete plans for his proposed Poor People's Campaign. At that time, there was a great deal of concern in white circles about the effect the campaign would have on Washington's poor blacks. Many city and federal officials were claiming that Stokely and other "black militants" would use the campaign to foment rebellions.

Stokely told us that he assured Dr. King that he would never do anything to embarrass him or subvert the campaign. He also promised Dr. King that he would do everything within his power to keep the people in Washington's black ghetto calm during the course of the campaign. Though Stokely did not believe that the campaign would be productive, he was committed to doing whatever he could to help Dr. King—because he loved and respected him.

Friday morning's newspapers carried extravagant accounts of the damage the night before. More than 200 stores had broken windows; 150 stores looted; 7 fires; more than 200 arrests; 35 injured and 1 white man was allegedly killed by a group of black youths during a fight involving him and 4 white companions.

On Friday afternoon, Stokely and Lester McKinnie, a member of the Washington SNCC staff, held the previously planned press conference. Stokely, who did most of the talking, was still very distressed by Dr. King's death. The reporters were very quiet while he talked.

"You may or may not know that this press conference was called before Dr. King's murder. We called it then to deal with Brother Rap Brown because we were very upset. Brother Rap Brown had been in jail for forty-one days and Governor Agnew of Maryland still seems to persist with his nonsensical charges so the brother can't get out of jail. We want him out next week when his trial comes.

"As for Dr. King's murder, I think white America made its biggest mistake because she killed the one man of our race that this country's older generations, the militants and the revolutionaries and the masses of black people would still listen to.

"The rebellions that have been occurring around these cities and this country are just light stuff to what is about to happen.

"We have to retaliate for the deaths of our leaders. The execution for those deaths will not be in the courtrooms. They're going to be in the streets of the United States of America."

The questions asked by the reporters when Stokely finished speaking indicated that they had little real understanding of the anguish and despair brought on by Dr. King's death.

"What accomplishments or objectives do you visualize?" one reporter asked.

"The black man can't do nothing in this country," Stokely answered. "We're going to stand up on our feet and die like men. If that's our only act of manhood, then goddammit, we're going to die."

"One last question," another reporter asked. "Do you fear for your life?"

"The hell with my life!" Stokely said. "You should fear for yours. I know I'm going to die. I know I'm leaving."

The press conference ended on that note. It was bitter, but what more could be expected?

Immediately after the conference, Bill Hall and Mickey McGuire (two members of the Washington SNCC staff), Stokely and I went to Howard University. There was a rally scheduled and Stokely was supposed to give a speech. There were scores of black teen-agers milling about in the rubble-filled streets surrounding the campus. It was obvious that they were still angry. It was also obvious that the city was in for more of what it had gotten the night before.

Although the Howard administration had scheduled a memorial service at the same time as the rally, there was a sizable crowd in attendance. Many of those present wanted to raze the city.

"We oughtta burn the whole mothafucking town down," declared an angry student from the rear of the crowd.

Although I understood their feelings, I disagreed with those who advocated massive violence. I knew that the police would resort to machine guns and grenade launchers to keep control.

After about fifteen minutes, Stokely, who had been standing in the rear of the crowd with Bill, Mickey and me, moved forward to the speaker's platform. He cautioned the students against going into the streets unprepared.

"Stay off the streets if you don't have a gun," he warned. "There's going to be shooting."

But by late evening, the rebellion was on. The streets were teeming with people. Everything that had happened the night before seemed to have been magnified about five times. The police and fire sirens were unending. I began to get worried.

"Look, Stokely, we'd better get out of here," I told him. "Every cop in the city knows you live here. This is a perfect setup. They got Dr. King last night. You might be next. We've gotta get outta here."

Bill Hall agreed. Unfortunately, Stokely did not. He insisted that we remain in the community "with the people." We argued back and forth for several minutes and I finally relented. We spent a virtually sleepless night.

Late Saturday afternoon, we were returning to the apartment from the SNCC office when we were stopped by a police officer.

"Everybody out of the car!" he commanded.

"What is it now?" I asked.

"Up against the car," snapped the cop. "You're in violation of the curfew."

"What are you talking about? The curfew's not in effect yet," I replied.

"It started a few minutes ago. Four o'clock to be exact."

With that, the cop moved forward in an attempt to search us. He reached for Stokely first.

"Get your hand off me," yelled Stokely. "If you're going to arrest me, arrest me, but keep your hands off me."

This angered the cop, who had a phalanx of buddies stationed in the middle of the street less than a half-block away. While Bill and I were trying to calm Stokely, a lieutenant from the Washington police force stepped forward and took charge. He knew Stokely and obviously didn't want any trouble.

"Okay, so you didn't know the curfew was on. Get back in your car and go on home."

Bill and I grabbed Stokely by the arms and steered him into the car. He slid in behind the driver's seat. I sat beside him and Bill climbed into the rear. I knew that Stokely was very upset so I watched him closely. The car was headed in the direction of the officers standing in the street. After starting the car, he stepped on the gas and began to ease the gear shift into drive.

"Hey, man!" Bill and I yelled simultaneously. With that, Stokely put the car in reverse and backed slowly down the street.

When we got back to the apartment, I gave Stokely an ultimatum.

"Pack whatever you need for the next few days, we're getting out of here. It's too dangerous to remain any longer."

"I'm going to stay," he declared.

With that, Bill, Mickey and I picked up our coats and stalked out. When we got to the lobby, I stopped. "Look, we can't leave him. How will this sound? 'Stokely Carmichael found dead in his apartment. He was last seen by three SNCC members who left him earlier in the day because they were convinced that his apartment was too dangerous.'"

We agreed that we couldn't permit that to happen. Bill said that he wasn't going to return to the apartment if all we were going to do was argue. Mickey concurred.

"No more arguing," I told them. "He's coming with us, one way or another."

I knocked twice before Stokely answered the door.

"Okay, get your shit," I told him.

He burst out laughing. He knew what we had in mind and realized that additional protest was useless.

We used a borrowed car in order to shake the police, who were following us. Because he was better known than the rest of us, Stokely had to lie on the floor in the rear. We spent the night in a house owned by the Tanzanian government. It was located in an upper middle-class neighborhood far from the scene of the rebellion.

We remained in Washington for a couple of days until we could get our plans together. We wanted to attend Dr. King's funeral, but there were some problems involved, the main one being transportation. We didn't dare fly. Several politicians in and around Washington were claiming that Stokely was responsible for the rebellion that had virtually paralyzed the city. We were certain that he would be arrested, in much the same manner that I had been in Orangeburg, if he showed his face at an airport.

We finally decided that the best way to get to Atlanta was to drive. This was no easy thing to do. Several towns along the route had also been disrupted by rebellions and many of them were under martial law.

"All we need to do is be caught in some Southern possum jump after curfew," said Mickey.

We carefully monitored news reports of those cities where rebellions had occurred. We then mapped out a route. We left Washington early in the morning accompanied by Stokely's fiancée, Miriam Makeba. Luckily, the trip to Atlanta turned out to be uneventful.

After we arrived in Atlanta, we were faced with another problem—"How the hell are we going to get into the church to attend the service?" No one knew we were in town. And we couldn't get any calls through to the King family. After surveying the church and its environs, we worked out a plan.

We got up very early on the morning of the funeral and headed for the church. We had almost miscalculated. There were thousands of people in the streets. We hadn't expected them to be there so soon. We had to fight our way through the milling crowd for almost a half-mile before we were in position to implement our plan.

When the buzz of the crowd informed us that the King family was about to enter the church, we bunched up and got ready. We were stationed within three feet of the entrance. As soon as the family moved past, we pushed forward in the form of a wedge and plowed in behind them. Miriam was stationed in the middle of the wedge—behind me and between Stokely, Bill and Mickey. The burly guards stationed at the entrance were so surprised by our audacious act that they stepped back and let us in.

We sat a couple of rows behind the family, surrounded by dignitaries, most of them rich and white. Hubert Humphrey, our old nemesis from Atlantic City, had a prominent seat. Jackie Kennedy, who had never shown any particular interest in Dr. King or black oppression, was also conspicuously present. Unfortunately, there was no place in the church for the hundreds of thousands of poor blacks who loved Dr. King and believed in his beautiful "Dream."

The service was just starting when I saw a familiar figure step through a door in the rear of the church. It was Willie Ricks. I don't know how he got into the church or how long he had been hiding back there. But I was glad to see him. He was a little uncertain of himself until he saw us and took a seat nearby.

We were the only SNCC people who managed to attend the funeral.

CHAPTER 20

SNCC Is Dead

THE FUNERAL WIPED me out. I still don't remember much about it. Dr. King's recorded eulogy, the mournful spirituals, the weeping: these things remain in my mind. I also remember thinking about others who'd been cut down in the struggle—Medgar Evers, Herbert Lee, James Chaney, Andrew Goodman, Mickey Schwerner, Jimmie Lee Jackson, Wayne Yancey, Jonathan Daniels, Sammy Younge, Malcolm X. . . .

So many good men.

After the service, we filed out of the church into the bright sunshine. My eyes were filled with tears, my throat ached and my hands shook. Ricks and Stokely were in the same condition. Miriam tried to comfort us, but the sense of eternal loss was too great. After marching for about a half-mile behind the mud-splattered, mule-drawn wagon that carried the coffin, we decided to return to the motel.

There we could mourn in private.

Less than three weeks after the funeral, I returned to court to be sentenced for refusing the draft. I appeared before District

Court Judge Newell Edenfield. Because I was still upset about Dr. King's assassination and angry about the prospect of going to prison, I let it all hang out when Judge Edenfield asked me if I had anything to say before being sentenced.

"The only people who can sentence me are black people," I declared. "Therefore, the only thing I can say is that you are prepared to carry out whatever you are, and I will fight as the rest of the black brothers have for the liberation of black people . . . till my death, I will fight for that."

Judge Edenfield's face turned from ruddy pink to splotchy purple while I spoke. When I finished, Howard Moore attempted to soften my words by putting them into context. Judge Edenfield understood what I meant, however, and remained unmoved by Howard's comments. He gave me the maximum sentence—five years.

Up to that time, the customary sentence for persons convicted of refusing the draft in Atlanta District courts was three years. Observers suggested that I got two extra years because of the militant nature of my presentencing statement.

Although Howard informed Judge Edenfield that my conviction was to be appealed, he refused to grant my release on bail. I was led from the courtroom and brought to Atlanta's Fulton County Jail. After two days, I was taken to a second jail in Newnan, Georgia, a small country town located about seventy-five miles from Atlanta. They kept me in Newnan for two days before transferring me to a third jail in Rome, Georgia. I remained in the Rome jail for two and a half weeks before being transported to the federal prison in Tallahassee, Florida.

All this time I was kept in solitary confinement and not permitted to receive mail or visitors. I asked my captors why this was being done on several occasions, but they always refused to answer. The Florida prison officials were a bit more magnanimous; they placed me in a cell block with other prisoners—in the maximum-security section.

I didn't have any real trouble until a guard attempted to cut my hair.

"No," I declared. "I've seen several prisoners with hair longer than mine. I don't intend to have my hair cut."

I had a long Afro-natural and wanted to keep it. It was one of the few things that helped me resist the feeling of complete hopelessness that threatened to overcome me. Although he was surprised by my refusal, the guard did not press the issue. I felt real good the rest of the day.

Early the next morning, I was awakened by four guards.

"Okay, Sellers, it's time for that haircut." Realizing that it was useless to try to fight all four of them, I went along. By the time I got to the prison barber shop I was angry and refused to sit in the chair. The guards grabbed my arms and wrestled me into the chair. The four of them held me while an inmate barber quickly shaved my head.

Late that evening, the inmate who cut my hair was placed in a cell directly beneath mine. I later found out that he was put there because the brothers on the yard had threatened to do him in for cutting my hair. This news made me feel better.

The racial situation in the prison was horrible. Although I was never permitted to leave my cell, I received information regularly—via the grapevine. Much of the tension was generated by the prison administration's response to requests by black prisoners that they be permitted to honor Dr. King. All such requests were blithely ignored. This was true even though the white inmates were permitted to make and fly a Confederate flag after the death of Alabama's Lurleen Wallace.

Tensions were heightened after a group of white prisoners who had been convicted of bombing a black church were brought into the prison. Although they were outnumbered two to one, the black prisoners vowed to "get" the church bombers. The inevitable occurred late one Sunday afternoon when a minor argument in the yard turned into a huge brawl between blacks and whites.

I watched the fighting, which raged for about fifteen minutes, from the small window high on the wall in my cell. Things were so bad after the fight that the prison administration was forced to segregate the prisoners. All the blacks were placed in one barrack and all the whites in another. Staggered mealtimes were also initiated.

Shortly after the brawl I was shipped out of Tallahassee to the federal government's maximum-security prison in Terre Haute, Indiana. Thin-lipped federal marshals transported me in a car. During the entire trip, I was bound by leg, wrist and waist chains.

I was greeted in Terre Haute by a group of ill-tempered white guards.

"We hear you're a black agitator," one of them said.

"We don't like any of that black militant shit around here," another declared.

"We don't want no progress talk around here," said a third.

"If you step your black butt out of line one inch, you're going to belong to us," a fourth said.

I was detained in the prison hospital for two days until the warden found the time to "greet" me. He turned out to be a stern man with a military bearing and icy eyes. He had heard about the brawl in Tallahassee and was convinced for some reason that I was responsible. My comments to the contrary were ignored.

"We don't intend to have any of that kind of stuff around here," he admonished. "We have a good prison and we intend to keep it that way."

"What's a good prison?" I asked.

He didn't answer.

I'd be lying if I said I wasn't frightened. There was a lot of anger among the white guards and inmates about the Tallahassee brawl. They, too, were convinced that I was responsible. Moreover, they had heard about the charges against me in Orangeburg. I was afraid that an attempt was going to be made on my life. I was always careful whenever I had to leave my cell. I slept very lightly.

In late August, I was taken from the prison and transported back to Atlanta. Howard Moore, whom I wanted to kiss when I heard the news, had managed to arrange my release. He had had to go all the way to the Supreme Court to do it. Judge Edenfield had used my presentencing statement as grounds for turning down bail petitions on four occasions. Supreme Court Justice Hugo Black had ordered him to release me.

Justice Black, who discounted a lower court claim that I was "dangerous," ruled that I be released on an appeal bond of no more than $5,000.

I was taken to court on August 22, where my release bond was established at $2,500. After the money had been put up, I was taken to the federal marshal's office to be freed. Although dog-tired, I was very happy. I was looking forward to spending the night in a *real* bed, after having a meal of *real* food. I received a depressing surprise, however.

As soon as the federal marshal removed the handcuffs from my wrists, a deputy sheriff from Louisiana stepped forward and clamped on another pair. I was under arrest.

I spent the next two nights sleeping in a *jail* bed and eating *jail* food.

The Louisiana deputy arrested me for a concealed-weapons charge lodged against me in Baton Rouge. In 1967, Stokely and I had gone to Baton Rouge for a speaking engagement at Southern University. On the way to the airport afterward, the car in which I was riding was stopped by the police, who were obviously looking for Stokely. Fortunately, he had decided to return to the airport in another car.

When the police searched us, they found a gun. It belonged to Stokely, and we were held until he came down and talked with the police. After he assured them that he was leaving the state immediately, they released us. I hadn't heard anything more about the affair until the deputy went into his act in the Atlanta marshal's office.

Although it was obvious that the state of Louisiana didn't really have a case against me, I was not released until five hundred dollars was raised to cover the five-thousand-dollar bond for the concealed-weapons charge. Shortly thereafter, I had to go to Louisiana and stand trial. I was found guilty, given a six-month suspended sentence and placed on two years' probation. One of the terms of the probation was that I could not possess a weapon at any time for any reason—on my person or in my home.

By the time I walked out of that courtroom in Louisiana, I didn't know if I was going or coming. My life was in total disarray. Everything I owned was gone, I'd never gotten the opportunity to return to Orangeburg to claim my clothing or my papers. Sandy was living in Orangeburg, but I couldn't get anything straight out of her. I wrote several times asking her to try to find my clothing, but she never answered my letters.

If Howard Moore hadn't worked with me, tried to help me get myself together, I don't know what I would have done. With the exception of Willie Ricks, who permitted me to share his apartment in Atlanta, I had almost no contact with SNCC. This was not surprising because there was no one working in the National Office. From time to time, I got word that Jim Forman and a couple of others were attempting to keep the New York office open, but no one ever knew anything about what they were doing. I didn't have enough money or energy to go to New York and find out.

One of the most disturbing bits of information I received during this period was that Stokely had been fired from SNCC. I didn't believe it when I first heard it, but after checking into the matter, I found out that it was true. Stokely's firing was officially announced on August 22 by Phil Hutchings, who was apparently SNCC's new "national program secretary."

Hutchings was reported to have said when making the announcement that ". . . it has been apparent now for some time that SNCC and Stokely Carmichael were moving in different directions."

Hutchings also claimed that Stokely had been engaged in a "power struggle with another organization member and that during the week of July 22, this struggle almost resulted in physical harm to SNCC personnel and threatened the existence of the organization."

These were serious charges. And I spent much of my time during the next few weeks investigating them. As I suspected, the other organization to which Hutchings referred was the Black Panther party. The "power struggle" did occur, but not quite in the manner that Hutchings and several other sources reported.

It all began in February, 1968, when Stokely, Rap and Jim Forman traveled to Oakland, California, to speak at a birthday rally for Huey Newton. The rally, which attracted over five thousand people, was held in the Oakland Coliseum. Huey was in jail at the time, charged with killing an Oakland police officer in a shootout. The Panthers arranged the rally in order to generate support for his defense and to raise money for legal fees.

Eldridge Cleaver made a momentous announcement during the rally; Stokely, Jim and Rap had been inducted into the Black Panther party. Stokely was the honorary prime minister, Forman the minister of foreign affairs and Rap was to be the minister of justice. The crowd roared its enthusiastic approval. Cleaver also announced that SNCC and the Black Panther party had established an "alliance." This news precipitated another enthusiastic roar from the crowd.

Few of those who attended the rally knew that there was a great deal of tension between Jim, Rap and Stokely. As a matter of fact, Rap and Jim were hardly on speaking terms with Stokely. They were upset because of comments he had made during a recent world tour and because he had refused to permit them to censure him upon his return. The Panthers were aware of these tensions, but didn't believe that they would have any significant effect on the "alliance."

I've never been able to ascertain who came up with the idea for the SNCC-Panther alliance. It was obvious, however, that both organizations had something to gain by it. SNCC, which was virtually defunct, was given new life. The Panther party was young, vital and growing rapidly. Its leaders had the ear of militant young blacks in the ghettos, a group with which SNCC had never really managed to develop any significant rapport.

On the other hand, SNCC had a reputation. It also had the big names—Carmichael, Rap and Forman. The Panthers, who were desperately trying to get exposure for Huey, needed the attention that SNCC leaders could attract.

"What we have done is worked out a merger with SNCC,"

Cleaver announced in the March 16 issue of the Panther newspaper, *The Black Panther.* "The Black Panther Party for Self Defense and SNCC are going to merge into a functional organization."

This proved to be impossible, primarily because of Jim Forman. Jim, who had always been a decision-maker and a leader, wanted to gain control of the Panther party. He was intent on directing its members in much the same manner he had directed SNCC's. The Panthers, who obviously felt closer to Stokely than to Jim and Rap, knew nothing of this. As a matter of fact, they probably wouldn't have entered into the alliance had it not been for their admiration of Stokely.

In any event, Jim Forman had a nervous breakdown shortly after the alliance was announced. It was brought on by the tremendous strain he had been under while trying, almost single-handedly, to keep SNCC alive. He was not completely incapacitated by the breakdown, however. And after resting up, he continued to work. One of the major results of Jim's breakdown, which remained a close-kept secret, was paranoia. Afterward, he was *convinced* that he was being watched and that those around him were spies.

In mid-July, 1968, Jim got in touch with the Panthers and suggested a SNCC-Panther rally at the United Nations. He suggested that the rally could be used to emphasize the Panthers' demand for a United Nations plebiscite to determine the political destiny of black Americans. He told them that the rally could also be used to draw attention to the legal problems of Rap and Huey.

The Panthers, who were very worried that Huey was going to be railroaded into the gas chamber, thought that the rally was an excellent idea. They immediately sent a contingent from their Oakland, California, headquarters to New York to work out details for the proposed rally. Two of them met with Jim Forman in SNCC's New York office.

Jim, who was still suffering from extreme paranoia, refused to speak with the two Panthers. He insisted that they communicate with written notes. Although they didn't understand what was going on, the Panthers complied with his demand. At one point, Jim passed them a note. After they read it, he took it back and tore it up. Shortly thereafter, he left the room.

Upon returning, Jim demanded that the two thoroughly confused Panthers return the note. They told him that they had already done so, but he didn't believe them. He insisted that they remove all their clothing. After they complied, he proceeded to search their clothes for the already destroyed note.

A couple of days later, several Panther leaders, including Bobby Seale and Eldridge Cleaver, arrived in New York. When they were informed that nothing had been done to organize the rally, they became angry. After a quick conference with their advance men, the Panthers requested a meeting with Jim Forman and his staff.

If I am not mistaken, Jim's staff consisted at the time of Stanley Wise, Donald Stone and Bob Smith. He may have had a couple of other people working with him. In any event, none of them knew what the hell was going on. Jim hadn't told them anything about his meeting with the Panther advance men.

All eyes turned to Jim when the meeting began. He remained silent. As a matter of fact, he reportedly acted as if he didn't even know why the meeting had been called. After a few minutes, Bobby Seale took charge. He asked Jim to have a seat on a nearby desk and demanded that he explain his actions. The other SNCC members were wondering why the Panthers were so upset. The meeting ended with the Panthers threatening to extract physical retribution if Jim didn't get himself together for the press conference scheduled for the following morning.

Some reports of the angry meeting have alleged that guns were drawn, but they weren't. Everyone present was aware, however, of the imminent threat of guns being brought into play. This is especially true of the Panthers, who were acutely aware that Huey's life depended on them.

The greatest casualty of the meeting was the SNCC-Panther alliance. It had no chance of surviving after the disastrous events surrounding the United Nations rally. As the summer progressed, the two groups moved further and further apart. Stokely grew closer to the Panthers during this period. Although there were significant differences between his political ideas and those of the Panther party, he did identify very closely with them, certainly more than he did with SNCC. This is the real reason why he was drummed out of SNCC by Phil Hutchings and Jim Forman.

I remained in Atlanta through the summer and fall of 1968. Because there was no SNCC activity in the area, I spent a lot of time talking with students in the Atlanta University Center. They were intensely interested in Black Power, SNCC and the Orangeburg assassinations.

Each Tuesday afternoon, I had to go downtown to the federal marshal's office and sign in. This was one of the provisions of my release from prison. Whenever I wanted to leave town, say to visit my parents in Denmark, I had to get permission from the federal marshal—another provision of my release.

I considered the entire procedure oppressive and degrading, but there was nothing I could do. The government had me in an eighty-three-year bind—seventy-eight for Orangeburg and five for refusing the draft.

In late December, Jim Forman came to town to conduct a SNCC staff meeting. Ricks and I decided to attend. To our surprise, we didn't even know most of the fifty-odd people in attendance. Although they were acting as if they were old SNCC veterans, the majority had joined the organization after Stokely was fired. They were openly hostile to Ricks and me. We were still friends with Stokely. And he was considered persona non grata.

I remained at the meeting for one reason: I wanted to see what Jim Forman had in mind. It didn't take me long to find out. One of the guests of honor was Charles Koen, the leader of a group of militant blacks from St. Louis, Missouri. Koen's group, which was called "The Liberators," was somewhat similar to the Panthers. It was obvious that Jim intended to use The Liberators in much the same manner as he had attempted to use the Panthers.

At one point during the meeting, Ricks and I were put on the spot. Jim was trying to engineer a "merger" with The Liberators. They were a small, inexperienced group, however, and he needed someone to organize them. With the exception of Jim, Ricks and I had more organizing experience than anyone at the meeting. Those in attendance appointed us the task of organizing The Liberators. Neither of us was interested. We had no intention to go to Saint Louis. And we didn't agree with Jim's scheme. I was convinced that it was just another desperate pipedream.

When we refused to organize The Liberators, we were fired!

Fired from SNCC!

Fired by a group of posturing idiots! I looked at Ricks and saw what must have been reflected in my eyes: surprise and embarrassment. We didn't argue or protest. There was no need. The situation was too ridiculous for words.

I looked into Jim Forman's eyes. Although he looked quickly away, I saw that he, too, was embarrassed.

What have we been reduced to? I asked myself as Ricks and I walked slowly down the hall toward the paper-cluttered stairway that led outside.

I was sitting in the waiting room outside Howard Moore's

office the next afternoon waiting to see Howard when Charles Koen, two of his Liberators and one of Jim Forman's new recruits approached me.

"Hey, man. Can we talk to you?"

"Sure, come on in here," I replied as I rose to lead them into an adjoining room.

"Why are you trying to destroy SNCC?" one of them asked immediately after we got into the room.

"Look, you jive-ass niggers don't even know what the hell . . ."

Before I could finish the sentence, one of them struck me in the back of my head with his fist. Before I could react, all four of them grabbed me from the rear. Although they were beating me about the head, neck and shoulders, I managed to get my feet under me before they struck a serious blow.

Gathering my strength, I gave a big heave forward. The four of them tumbled to the floor in front of me. A huge black pistol fell out of one of their pockets and landed heavily on the floor.

Before they could react, I whipped out the .38 caliber pistol I carried in my back pocket, flicked off the safety and pointed it at them. Scrambling clumsily to their feet, they charged toward the door. I calmly raised my arm and tightened my finger on the trigger. I was deciding which one to shoot first when Howard Moore's voice cut through the noise.

"Cleve! Don't do it."

"What's going on?" he asked while they wrestled with each other in an attempt to get through the narrow doorway.

"I'm trying to decide which one of these niggers I oughtta shoot first."

"Don't do it," Howard said.

"Yeah, you're right," I replied, my anger already turning to sadness.

By this time, my *assailants* had disappeared through the door. I could hear their heavy footsteps receding rapidly down the hall.

"Who were those men?" Howard's secretary asked over his shoulder. "Were they members of SNCC?"

"No," I replied. "They weren't members of SNCC. SNCC is dead."

Yes. I knew that really was the end. SNCC was dead.

CHAPTER 21

I Have Only One Life: The Struggle

THE PAST FOUR years have been tough, very tough. The black liberation struggle has gone through many discouraging changes. We have enjoyed some minor victories, but for the most part we've continued to labor in a quagmire of problems, uncertainties and systematic oppression. Many of those who stood with us in the early years are no longer present. Some lost strength, others lost perspective and far too many lost their lives. For those who remain, the struggle has become a way of life; that which defines purpose, aspirations and existence.

The sense of common cause and shared destiny that kept the struggle functionally coherent up through the midsixties is gone. The movement is fragmented into localized groups primarily responsive to community and regional interests. The assassination of Dr. King and the collapse of SNCC have left a tremendous void that no individual or organization has managed to fill.

Black Consciousness remains strong in most sections of the

nation, the East Coast in particular, but those who possess it generally suffer a lack of direction. The black masses are no longer caught up in the drama and promise of the movement. They are passive, waiting on a viable vision, a new concept of success. The present state of confusion can be traced almost directly to 1968, the year that signaled the end of the ascendant Black Power phase of the movement. By the latter portion of that year, the black movement was in the process of dividing into two factions: Cultural and Political Nationalists.

SNCC's basic Black Power analysis was the ideological foundation for both factions. Each was based on the premise that American blacks are colonized and in need of unity, power and a coherent racial identity. Both factions also accepted the premise that black oppression cannot be eliminated without a full-scale revolution, probably a violent one. There were several important differences between the two factions, however.

The Cultural Nationalists believed that blacks had to get themselves together culturally before engaging in overt attacks on the system. They turned to Africa and adopted certain symbols that they considered representative of "negritude." They favored African names, hair styles, family organization and art. They opened small shops in nearly every major city in the nation during 1968 and 1969, which sold a wide variety of items imported from Africa.

The Cultural Nationalists were deep into African history. They felt that ancient Africans utilized many group-organization and decision-making techniques that blacks should emulate. One of the concepts that many Cultural Nationalist groups strongly supported was *Ujama,* a term that meant African Socialism. Although the practice was not restricted to them, many Cultural Nationalists stopped using their American names and adopted African or Arabic ones.

One of the most prominent Cultural Nationalist groups was headed by Ron Karenga. His group, which was headquartered in Los Angeles, was called US. Karenga, who sported a striking Fu Manchu moustache and a bald head, was the archetypical Cultural Nationalist. Members of his organization wore dazzling African-style robes and spent much time attempting to organize young Los Angeles blacks by the use and manipulation of African cultural symbols.

In an article printed in the magazine *Negro Digest* in January, 1968, Karenga presented some of his ideas about the goals of Cultural Nationalism. "We have said and continue to say, that the battle we are waging now is the battle for the minds of Black people, and that if we lose this battle, we cannot win the violent one."

Like all Cultural Nationalists, Karenga and his followers spent much time attempting to define the revolutionary function of art. "Black art," he claimed, "must expose the enemy, praise the people and support the revolution. It must be like LeRoi Jones' poems that are assassins, poems that kill and shoot guns and 'wrassle cops in alleys taking their weapons, leaving them dead with their tongues pulled out and sent to Ireland.'"

In order to get his ideas directly to the people, Ron Karenga wrote a small book titled *The Quotable Karenga.* It contained his theories and was similar to Mao Tse-tung's "little red book." I don't think it ever got to be a best seller, in the black or white community.

Political Nationalists were generally scornful of their Cultural counterparts. They were convinced that the great amount of attention being lavished on art and culture was dysfunctional and misleading. They were in favor of armed confrontation. They frequently contended that "the obligation of the revolutionary is to make revolution." They also said that the most revolutionary act possible at the time was taking up a gun and going down on the "pigs."

The Black Panther party was, of course, the most prominent exponent of Political Nationalism. The party possessed a hodgepodge ideology jerry-built from numerous sources; Mao Tse-tung, SNCC, Marcus Garvey, Fidel Castro, North Korean Communists, Frantz Fanon and Karl Marx. They generally called themselves Marxist-Leninists.

The Panthers adamantly claimed that they had managed to bridge the age-old ideological chasm between conventional Marxist analyses of class oppression and traditional Black Nationalist analyses of racial oppression. They claimed that their ideology permitted them to speak to race and class oppression at the same time. They called themselves "The Vanguard" and acted most of the time as if the struggle for black liberation did not officially begin until Huey Newton and Bobby Seale founded the party.

Unlike the Cultural Nationalists, who contended that armed warfare should come *after* the people were educated and united, the Panthers and other Political Nationalists advocated immediate confrontation. The symbols of their organizations were the tools of warfare: rifles, bombs, bandoliers and pistols. Their weekly newspaper, *The Black Panther,* carried detailed instructions on the care and use of all kinds of guns. The militant cartoons in the newspaper all but instructed readers to go out and start "offing pigs."

Political Nationalists spent much time and energy haranguing the police. They claimed that the elimination of police violence in black communities was the first step toward liberation and self-determination. In March, 1968, the Panther newspaper carried a terse warning to all police officers "bent on killing black people":

> HALT IN THE NAME OF HUMANITY! YOU SHALL MAKE NO MORE WAR ON UNARMED PEOPLE. YOU WILL NOT KILL ANOTHER BLACK PERSON AND WALK THE STREETS OF THE BLACK COMMUNITY TO GLOAT ABOUT IT AND SNEER AT THE DEFENSELESS RELATIVES OF YOUR VICTIMS. FROM NOW ON, WHEN YOU MURDER A BLACK PERSON IN THIS BABYLON OR BABYLONS, YOU MAY AS WELL GIVE IT UP BECAUSE WE WILL GET YOUR ASS AND GOD CAN'T HIDE YOU.

Despite the Panthers' impressive plans and considerable arrogance, the masses of blacks were not particularly turned on by the Political and Cultural Nationalists. Neither faction received the same broad support that was lavished on SNCC, CORE, SCLC, the NAACP and the Urban League during the initial phases of the movement.

The schism between the Cultural and Political Nationalists, which led to armed confrontations in some cities by the latter portion of 1969, was overshadowed during most of that year by government repression. In a very real sense, 1968 witnessed warfare between black revolutionaries and various militaristic arms of the government. Nearly every militant black group in the nation was under attack by squads of death-dealing "law-enforcement officers."

I am convinced that President Nixon, his flunky John Mitchell and the late FBI Director J. Edgar Hoover were largely responsible. Nixon ran for the presidency on a barely disguised repress-niggers-and-other-malcontents platform and the streets ran wet with blood immediately after he took office. In city after city, cadres of helmeted police officers launched vicious attacks on blacks. The Panthers, whom Mr. Hoover accused of being a group of organized hoodlums, received the brunt of the attack. They lost several members in shoot-outs with police.

The government was obviously engaged in a systematic attempt to intimidate the black liberation struggle. The heat was on and every prominent black revolutionary was feeling it. It seemed that the government's leaders would not be satisfied until we were silenced, imprisoned, forced to flee the country or killed.

Rap Brown had fourteen criminal counts in four states and the District of Columbia threatening his future as a free man. At the same time he was appealing a five-year sentence imposed following conviction on an obscure federal gun charge in New Orleans. Stokely was faced with a ten-year prison term as a result of charges stemming from the inconsequential Atlanta rebellion. Phil Hutchings, who had replaced Rap as SNCC's chairman, was faced with a trial in Saint Louis.

Several former SNCC members were embroiled in legal difficulties in Texas. One of them, Lee Otis Johnson, was sentenced to thirty years in prison for allegedly selling one marijuana cigarette. In Houston, Texas, five militant Texas Southern University students were faced with murder trials in connection with the death of a policeman during a police raid on their campus. The five were being tried even though prosecutors admitted that most of them were nowhere near the scene when the officer was slain.

Former SNCC member Fred Brooks of Nashville, Tennessee, was sentenced to five years in prison for violating a Selective Service law. Willie Ricks was facing conviction on trumped-up riot charges in Atlanta. The FBI was hunting Eldridge Cleaver for a parole violation. Huey Newton was in prison for allegedly killing a police officer. And Bobby Seale was under restrictive probation.

I could add more names and circumstances to the list, but I think I have made my point. The government was on the offensive and everyone who had taken a revolutionary leadership position seemed to be fair game. It was possible for the government to get away with this because the people were not in motion, not actively caught up in the struggle. I am not attempting to imply that all those who came under police attack were innocent. Some were guilty of the things charged against them. But most weren't. All that most of us had done was insist that black people had the right to self-defense and freedom "by any means necessary."

The government was hot on my trail and it was impossible for me to forget this. I felt suspended most of the time between the coffin and the jail cell. It seemed to be just a matter of time before I would end up in one or the other. Undercover agents followed me everywhere I went. Some of them became so familiar to me that I knew them. My phone was tapped. I had a five-year conviction for refusing the draft and it seemed that I was going to get a fifty- to sixty-year sentence as soon as South Carolina's officials could get me into court. There was a phrase popular in the black community at the time that seemed particularly applicable to my dilemma: "You can run, but you can't hide."

The weekly trip I was forced to make to the federal marshal's office in downtown Atlanta didn't help my frame of mind. It was one of the most degrading things I have ever been forced to do. I tried never to go by myself, but sometimes I had to. In an attempt to harass and keep me off balance, officials in the office frequently divulged bits of information they had collected on my activities.

I didn't have a job and lived an uncertain hand-to-mouth existence. I resided in a small roach-filled apartment in southwest Atlanta with Willie Ricks. We didn't really have any furniture, just a few threadbare items we'd managed to scrounge from friends and acquaintances. The electricity was never on because we couldn't afford to pay the bill. On those rare occasions when we wanted to hear a little music from Ricks's radio, we would jimmy the fuse box with wires and bypass the cutoff switch. One of our most difficult tasks was coming up with seventy dollars a month for the rent. We were always two and three months behind. And more times than I care to recall we had to hide behind the curtains so that the landlord wouldn't know we were home.

I am an early riser and my first stop after getting out of bed each morning was the refrigerator. I would do this in order to see if Ricks had managed to procure any food the night before. Usually he hadn't. After getting dressed, I would head out into the streets—to hustle. I never knew when I left if I would return that night, two weeks later or never.

All my days were essentially the same; trying to get food, looking for money-making opportunities and attempting to raise the general level of political consciousness of people in the black community. I was extremely embarrassed when I initially had to resort to asking people for food money, but I got over those feelings fast. Hunger helps you adjust.

I spent a lot of time working with students in the Atlanta University Center. Many of them were very interested in radical black politics. They were the source of great inspiration to me. Although they hadn't been down the roads I had traveled, they were sensitive to my experiences and were eager to learn whatever I could teach. Mostly, I tried to get them to understand the need for following through on the basic Black Power analysis developed by SNCC.

I urged them to buy guns and arm themselves. Always on the move, I repeatedly asked the same questions: "Do you have a gun? Do you know what the man is doing? They are preparing internment camps for those who speak out and demand their rights.

What are you prepared to do when they break down your door? Can you defend yourself?"

Early in 1969, I got an offer to teach a one-semester course on Black Ideology in the Black Studies Program at Cornell University. Although I didn't want to leave Atlanta, I accepted the job. My attorney, Howard Moore, thought it would be a good experience for me. He said that it would provide me with an opportunity to get some rest, get my thoughts together and acquire credentials that might favorably affect my legal situation. Of course, I had to get permission from the federal marshal before accepting the job and leaving town.

The course at Cornell, which lasted from January through June, was interesting. Teaching it was very helpful to my thinking. Unfortunately, I didn't get much rest or credentials that I dared present in court. Cornell was a hotbed of white racism, liberal confusion and black militance. Although I was not directly responsible, the campus got more national attention during that semester than it had in years. The dramatic picture of black students striding "resolutely" across the campus with rifles and shotguns accurately portrayed the incredibly hostile climate.

While working at Cornell, I made contact with blacks at Harvard University. Harvard's School of Education was recruiting blacks and I was urged to apply. I didn't expect to get accepted because I didn't possess a B.A., but I was. I spent the 1969–70 school year working on a master's degree in education at Harvard. The primary objectives of Harvard's black students were essentially the same as those which precipitated the crisis at Cornell: they were intent on controlling the structure, content and direction of their education. There were several student-administration confrontations during the year, but none as serious as the one at Cornell.

On May 4, 1969, Jim Forman emerged dramatically into the national limelight with an audacious gambit at New York City's famed Riverside Church. After demanding that the minister cut Sunday morning worship short, Forman read a list of demands for "reparations" to the congregation. Standing before confused listeners, Jim displayed the same nerve and verve that had made him such a dominant figure in SNCC.

He demanded rent-free office space in the church for a new organization he was tutoring, the National Black Economic Development Conference. He also wanted unrestricted use of the church's telephone, radio station and classrooms. Without a smile, he told them that he wanted 60 percent of the church's income from its stock and real estate holdings. Those were just the things he wanted from Riverside Church. Before finishing his unprecedented address, Jim informed his listeners that he wanted them to use Riverside's "influence and historic reputation" to pressure the nation's all-white Christian churches and Jewish synagogues to provide his organization with $500 million.

While a few of his astounded listeners filed out in outrage and others wept, Jim Forman explained some of the reasons for his demands. "Six million Jews were killed in Germany and Israel is still getting reparations. Fifty million blacks died in slavery and the black people have been paid nothing." Before leaving, Jim announced that the church had one week to respond to his demands.

Jim's action at Riverside was consistent with the Socialist direction in which he and many other black revolutionaries were moving. His comments to a group of students at Detroit's Wayne State University a week earlier encapsulated his analysis. "We must commit ourselves to a society where the total means of production are taken from the rich people and placed in the hands of the state for the welfare of all the people. This is what we mean when we say control.

"We are dedicated to building a socialist society inside the United States where the total means of production and distribution are in the hands of the state and that must be led by black people, by revolutionary blacks. Only by armed, well-disciplined, black-controlled government can we insure the stamping out of racism in this country."

Although I admired Jim's nerve, I was convinced from the beginning that he would not succeed. There were too many factors working against him, the most important one being his lack of unified support. There was no organization behind him. The Black Economic Development Conference was a facade; there weren't any warm bodies behind it.

In order to muster the "clout" necessary to make such demands and have them met one must have the unquestioned force of people, thousands of people, aligned in support positions. Jim was obviously bluffing. Jim Forman is no fool, however, and I am certain that he was aware of all this. I suspect that his real objective was publicity. He was probably using Riverside Church as a forum, a place to spread his Socialist ideas.

Jim Forman's gambit at Riverside Church was representative of the desperate tactics many black revolutionaries have been utilizing during the past four years. In the absence of a unified movement, individuals and groups have resorted to a plethora of schemes designed to "get something big going." Some have managed to win a modicum of support, but none have remained in the limelight for more than a few weeks or months.

The Black Panthers are a glaring exception. The party has gained increasing prominence during the past four years, but its program leaves much to be desired. Even though the Panthers have won respect from blacks who admire their courage under fire, the party is not actively supported by the black masses. This is so simply because the party has not come up with solutions to the problems of the black masses. Free clothing and breakfast programs are a step in the right direction, but they are only Band-Aid solutions. They cannot and will not lead to the necessary redistribution of wealth and power.

On March 10, 1970, Rap Brown was scheduled to appear in court in Bel Air, Maryland, to be tried on charges stemming from the controversial speech he made in Cambridge three years earlier. He didn't show. Immediate concern about his absence was overshadowed by a major tragedy that took place on the same day less than one mile from the courthouse where he was to be tried. William Herman (Che) Payne and Ralph Featherstone were blown to bits when a bomb went off in the car in which they were riding.

Maryland law-enforcement officials charged that the bomb was being transported by Che and Ralph and that it went off before they wanted it to. I have never believed this. The people with whom Ralph and Che worked in nearby Washington, D.C., didn't believe it either. They issued a statement charging that the two had been assassinated: "The presence of Ralph and Che in Bel Air was certainly known by their enemies. Also known was the fact that Ralph and Che were to be responsible for transporting Rap Brown safely into Bel Air. *A cold and calculated decision was taken to eliminate them.* During the night a high explosive bomb was placed in their car. . . ."

I heard about the assassination of Ralph and Che several hours afterward. I didn't ask any questions. I didn't say anything as a matter of fact. What was there to say? Friend after friend after friend, destroyed and obliterated. For what? Simply because they sought freedom and dignity. I was beyond crying.

All I could do was hurt—and try to quell the fathomless rage welling up inside me.

In late September, 1970, I had to return to Orangeburg to be tried for charges stemming out of the 1968 massacre at South Carolina State. Although innocent, I was still worried. My concern increased when I found out that city officials had ordered in one hundred troops from the South Carolina National Guard. The troops, who brought armored personnel carriers with them, had orders to remain in town for the duration of my trial.

There was a great deal of tension in the air when the bailiff called, "Hear ye, hear ye," announcing the beginning of the trial. I had two attorneys, Howard Moore from Atlanta and F. Henderson Moore from Charleston. They were grim, not expecting anything but a guilty verdict, which they hoped to have overturned in a later appeal.

Just before the judge entered the courtroom a contingent of gray-eyed highway patrolmen strode in and took seats along the entire front row of the courtroom. Behind them were seated a large number of black youths, some of them attired in African dashikis. Several of them wore small "Free Cleve Now" buttons. Judge John Gimball admonished those wearing the buttons as soon as he got into the courtroom.

"We see nothing wrong with the badges," protested one of my attorneys.

"Well I do," retorted Judge Gimball. "There's plenty wrong with it. It will probably influence the jury adversely to your client."

"We can see no difference in that kind of badge and the influence that might be exerted by having the courtroom and the courthouse filled with highway patrolmen," replied Henderson Moore. Judge Gimball still disagreed. He ordered everyone wearing one of the buttons to take it off or leave the room. After leaving the courtroom to remove the buttons, the young blacks filed slowly back and retook their seats. The tension increased.

I expected the state's case against me to be much stronger than it was. Of the ten who testified against me—all of them were connected with law enforcement—only one could cite an incident where I was observed breaking a law. Chief J. P. Strom of the State Law Enforcement Division testified that I "refused to disperse immediately when ordered." Orangeburg Police Chief Roger E. Poston testified that he saw me "move from group to group" during the disturbance but didn't hear me say anything. Under cross-examination by Howard Moore he admitted that he didn't see me commit any illegal acts. He also admitted that he had arbitrarily chosen to enforce the state's trespass statute while ignoring anti-discriminatory laws.

Less than a week after the trial began the jury of nine whites and three blacks returned their verdict. They found me guilty of the only charge left standing, refusal to disperse immediately when ordered to do so. Judge Gimball gave me the maximum sentence for the offense: one year in prison and a $250 fine. Anticipating an appeal, he set a $5,000 bond. He also granted me permission to travel out of the state when I agreed to waive extradition back to South Carolina should the sentence be upheld in the courts.

I have worked in several states on a number of projects during the past year and a half. For a time, I lent my efforts to Malcolm X University, an alternative college operated by blacks in Durham, North Carolina. I have served as a consultant and organizer for the militant Student Organization for Black Unity. I have also served as a coordinator for the African Liberation Day Committee. My objective has always been the same, helping black people get it together.

I am very tired. With the exception of time spent in jails and prisons, I haven't had a vacation in twelve years. I am always on the job, day and night, summer and winter. I almost never get paid, but I survive. I am sometimes hungry, but never for very long. When I need shoes or clothing, I get them—"by any means necessary."

I have spent a great deal of time studying revolutionary movements and political theory. All my closest associates do the same. Stokely has been living in Africa for the past several years, studying with Kwame Nkrumah, former premier of Ghana. On those few occasions when he returns to this country for speaking tours, I have gotten together with him for long rap sessions. Both of us are deep into Pan-Africanism. I consider myself a Pan-Africanist. I am convinced that the destiny of blacks in this country is inextricably linked to Africa.

I am committed to doing whatever I can to build bridges between blacks in this country and revolutionary groups in Africa. Africa has the people, the resources and the power potential to become one of the most dominant powers in the world. American blacks are Africans in exile. When Africa is free, her sons and daughters, wherever they are in the world, will have a solid base for their struggles.

I don't have a personal life anymore. I don't know where Sandy is or what she is doing. I sometimes want to drop everything and search her out, attempt to take up where we left off. It wouldn't work. I know that. Everything I have, all my strength, is still wrapped up in the movement. Things would turn out exactly as they did before.

I am caught up in the strong current of a river of no return. My being is inseparable from the struggle. I have thought about all this a great deal, who I am, where I've come from and where I'm going. I don't expect to live a normal life span. Nor do I expect to die a normal death. I am not unique. It's the same for almost everyone who *lived* the SNCC experience. Stokely, Rap, Jim Forman, Bob Moses, wherever he is, we have become one with the struggle. It doesn't matter that we share disagreements and petty animosities from time to time. We all want the same thing. We are all driven by the same inexorable force.

We will not stop because we can't. I take solace in the face of the hardships before me because I know that we are right and those who oppress us and our people are wrong. Most important, I believe that Dr. King was eminently correct when he said, "The arc of the moral universe is long, but it bends toward justice."

INDEX